Practicing Qualitative Methods in Health Geographies

Health geographers are increasingly turning to a diverse range of interpretative methodologies to explore the complexities of health and illness in relation to space and place in order to better understand and improve health care policy. Many health geographers are increasingly drawing on postmodernism and feminist health geography to explore issues of representation and the body to investigate the metaphorical, physical and emotional challenges of the body and disease.

Reflecting these interests, this comprehensive book explores a wide range of creative qualitative methods used to explore the psychosocial experiences of health care patients. This includes traditional methods such as in-depth interviews, focus groups, participant observation, diary studies and discourse analysis, as well as novel techniques such as 'go-along interviews', reflexive writing, illustrations and photographic techniques. Various qualitative research techniques unique to geographers are explored in depth, including the health and place approach, comparative case study analysis and the qualitative approach to the use of geographic information systems (GIS).

Illustrating how a diverse range of qualitative methodologies are used in a variety of health contexts, this book is a major contribution to health geography and will be of interest to both practitioners and academics alike, as well those with broader interest in qualitative research methods.

Nancy E. Fenton is Adjunct Professor in the School of Public Health and Health Systems at the University of Waterloo, Canada. She is involved in interdisciplinary qualitative health research investigating the relationship between the environment and health as it relates to risk perception, particularly among children and youth.

Jamie Baxter is Associate Professor in the Department of Geography at Western University, Canada. His research interests include: the social construction of risks from technological hazards, community responses to hazards, environment and health, noxious facility siting and social science research methodology.

Geographies of Health

Series Editors
Allison Williams, associate professor, School of Geography and Earth Sciences, McMaster University, Canada
Susan Elliott, professor, Department of Geography and Environmental Management and School of Public Health and Health Systems, University of Waterloo, Canada

There is growing interest in the geographies of health and a continued interest in what has more traditionally been labeled medical geography. the traditional focus of 'medical geography' on areas such as disease ecology, health service provision and disease mapping (all of which continue to reflect a mainly quantitative approach to inquiry) has evolved to a focus on a broader, theoretically informed epistemology of health geographies in an expanded international reach. as a result, we now find this subdiscipline characterized by a strongly theoretically-informed research agenda, embracing a range of methods (quantitative; qualitative and the integration of the two) of inquiry concerned with questions of: risk; representation and meaning; inequality and power; culture and difference, among others. Health mapping and modeling has simultaneously been strengthened by the technical advances made in multilevel modeling, advanced spatial analytic methods and GIS, while further engaging in questions related to health inequalities, population health and environmental degradation.

This series publishes superior quality research monographs and edited collections representing contemporary applications in the field; this encompasses original research as well as advances in methods, techniques and theories. The *Geographies of Health* series will capture the interest of a broad body of scholars, within the social sciences, the health sciences and beyond.

Also in the series

The Afterlives of the Psychiatric Asylum
Recycling Concepts, Sites and Memories
Edited by Graham Moon, Robin Kearns and Alun Joseph

Geographies of Health and Development
Edited by Isaac Luginaah and Rachel Bezner Kerr

Soundscapes of Wellbeing in Popular Music
Gavin J. Andrews, Paul Kingsbury and Robin Kearns

Mobilities and Health
Anthony C. Gatrell

Practicing Qualitative Methods in Health Geographies

Edited by Nancy E. Fenton
and Jamie Baxter

Routledge
Taylor & Francis Group
LONDON AND NEW YORK

First published 2016
by Routledge
2 Park Square, Milton Park, Abingdon, Oxon OX14 4RN

and by Routledge
711 Third Avenue, New York, NY 10017

Routledge is an imprint of the Taylor & Francis Group, an informa business

British Library Cataloguing in Publication Data
A catalogue record for this book is available from the British Library

Library of Congress Cataloging in Publication Data
Names: Fenton, Nancy E., author. | Baxter, Jamie, author.
Title: Practicing qualitative methods in health geographies / by Nancy E. Fenton and Jamie Baxter.
Description: Farnham, Surrey ; Burlington, VT : Ashgate, [2016] | Series: Geographies of health series | Includes bibliographical references and index.
Identifiers: LCCN 2015043321 (print) | LCCN 2016006525 (ebook)
Subjects: LCSH: Medical geography—Methodology. | Qualitative research.
Classification: LCC RA792 .F46 2016 (print) | LCC RA792 (ebook) | DDC 614.4/2072—dc23
LC record available at http://lccn.loc.gov/2015043321

ISBN: 978-1-4724-4539-1 (hbk)
ISBN: 978-1-3156-0194-6 (ebk)

Typeset in Times New Roman
by Swales & Willis Ltd, Exeter, Devon, UK

MIX
Paper from responsible sources
FSC® C013985
FSC www.fsc.org

Printed in the United Kingdom
by Henry Ling Limited

Contents

Figures

Tables

Contributors

Gavin J. Andrews is a Professor in the Department of Health, Aging and Society at McMaster University. As a health geographer his wide-ranging empirical interests include aging, holistic medicine, health care work, fitness and music. Most recently his research has engaged non-representational theory, and specifically the concept of the 'affect', to help rethinking the fundamental and immediate nature of human well-being. Much of his research is positional and considers the development and future of health geography. His publications include five edited books and over 120 journal articles and chapters.

Jamie Baxter has research interests that include social science methodology, geography of health, environment and health, sustainable energy, environmental risks from hazards, community responses to hazards, environmental justice and noxious facility siting. He has studied communities living with municipal and hazardous waste sites, urban pesticides and wind turbines. His research tends to be based on empirical research involving qualitative and mixed methods designs. He has co-authored a number of articles on qualitative research including pieces on case studies, interviewing, mixed methods and qualitative rigor.

Victoria Casey completed her MA in health geography at Simon Fraser University (BC, Canada) in 2013. Her thesis research focused on the practice of informal caregiving in medical tourism. She has published her research in journals such as *Globalization & Health* and the *International Journal for Equity in Health.*

Heather Castleden is a Canada Research Chair in Reconciling Relations for Health, Environments and Communities and is an Associate Professor in the Departments of Geography and Public Health Sciences at Queen's University. As a white settler scholar ally with training in health geography, she specializes in fostering healthier Indigenous-settler relations across a number of research settings. Heather has spent over 15 years developing partnerships with Indigenous peoples in Canada that align with their priorities and her thematic expertise: health equity through a social/environmental justice lens and the nexus of culture, place, power and ethics. She holds degrees from the

University of Manitoba (BA, native studies) and the University of Alberta (MEd, educational policy studies, and PhD, human geography).

Stephanie E. Coen is a PhD candidate in geography at Queen's University, Kingston, Ontario, Canada. She earned her BA (honors) and MA in geography from McGill University. Her doctoral research, supported by a Frederick Banting and Charles Best Canada Graduate Scholarship from the Canadian Institutes of Health Research, examines the influences of gender on barriers to and facilitators of exercise participation in gym environments. She is a member of the Campbell and Cochrane Collaboration-affiliated Sex/Gender Methods Group (http://equity.cochrane.org/sex-and-gender-analysis). Her research interests include gender and health, social contexts of health behavior, critical geographies of health and feminist and qualitative methodologies.

Tara Coleman has a PhD in human geography from the University of Auckland where she is a Professional Teaching Fellow in the Social Science for Public Health programme. Tara's PhD thesis explored the interrelations between 'place', 'being aged' and 'well-being' that shape the experience of aging-in-place for seniors living on Waiheke Island, New Zealand. Her MA in human geography considered school, home and Teen Parent Units as sites for sexuality education. Tara's research interests also encompass interdisciplinary approaches to health and well-being, feminist geographies, phenomenology, health policy and qualitative research methods.

Valorie A. Crooks is Associate Professor in the Department of Geography at Simon Fraser University (BC, Canada). She is also a Scholar of the Michael Smith Foundation for Health Research and Canada Research Chair in Health Service Geographies. She specializes in non-hypothesis testing, qualitative health services research and leads the SFU Medical Tourism Research Group (www.sfu.ca/medicaltourism).

Jennifer Dean is a social scientist with a PhD in health geography from McMaster University. Dr. Dean is currently an assistant professor in the School of Planning at the University of Waterloo (Canada) where her research program broadly focuses on social and environmental determinants of health, planning healthy and inclusive communities, and social equity and justice. She has a strong background in qualitative research methods, community-based research and has worked with various marginalized populations, including youth, recent immigrants and low-income families.

Eric Drass is an artist and curator who makes work in a range of media, from painting to digital installation to generative experiments which live on the net. Some of his favorite themes are identity, consciousness, the philosophical ramifications of artificial intelligence, big data and the relationship between humans and machines. Sometimes this work is political, frequently it is playful, and often it is provocative or transgressive in some way. His works are frequently reported and cited online (BoingBoing, b3ta, Imperica, Computer

Arts, Wikipedia, etc.). Eric holds a degree in philosophy and psychology (Oxford) and is co-author on a number of patents dealing with PRISM-type surveillance technologies (long before PRISM became public) and a number of academic papers relating to neural network models of language acquisition and heritability. He also used to be a singer in an experimental hard-core band, an unsuccessful male model and once took a dotcom 1.0 company from a bedroom project to 14 countries and back, spending US$50 million on the way.

S. Michelle Driedger is a Professor and Tier 2 Canada Research Chair in Environment and Health Risk Communication, in the Department of Community Health Sciences, with an affiliated academic appointment in the Department of Environment and Geography, at the University of Manitoba. A health geographer by training, her broad areas of research interests include public and health risk communication, risk perception and knowledge translation under conditions of uncertainty. Drawing primarily on qualitative methods, her research focus involves the study of how new and emerging risk and public health controversies develop in science, policy and public forums. She is particularly interested in how public health risk communicators can meaningfully engage public audiences to enable informed decisions about risk recommendations, including protective behaviors that can be adopted for individuals and their families. Her research explores these aspects with both general population and Metis contexts using a variety of data sources: interviews, focus groups, documents (e.g., policy, reports), traditional news media and more recently, social media.

Nancy E. Fenton is a social scientist with a PhD in educational policy and a Postdoctoral Fellowship in health geography. Dr. Fenton is an Adjunct Professor in the School of Public Health and Health Systems at the University of Waterloo involved in interdisciplinary qualitative health research investigating the relationship between the environment and health as it relates to risk perception, particularly among children and youth. She has authored and co-authored papers on health policy, children's health, child-centered methodologies and asthma and associated allergies in First Nations and Inuit communities in Canada. She is currently a research consultant in the McMaster Institute for Innovation and Excellence in Teaching and Learning at McMaster University.

Aaron Franks is a critical human geographer, theatre artist and multidisciplinary Postdoctoral Fellow at Queen's University. He is interested in littoral zones where practice and relationships churn up theory, and vice versa. Aaron has done collaborative research involving climate justice, the relationship between social movements and communities, environmental justice and health, and human rights in the context of settler colonialism and capitalist exploitation. He is a member of the Centre for Environmental Health Equity and the Health, Environment and Communities Research Lab. He holds degrees from the

University of Alberta (BFA in theatre), Brock (MA in social justice and equity studies) and Glasgow (PhD in human geography).

Theresa Garvin is Professor of Human Geography and Planning at the University of Alberta. Her research interests center on the health and social impacts of built and natural environments. These impacts can include the tension between people and the natural environment, such as the community-based impacts of oil and gas development. They can also examine the interactions between people and social environments such the challenges faced by seniors living in the suburban regions. In all her work, Dr. Garvin employs interdisciplinary perspectives and critical social theories of place to create and foster linkages between science, policy and civil society.

Robin Kearns is Professor of Geography in the School of Environment at the University of Auckland. He has a long-standing interest in the links between health and place and is an Associate Editor of the journal of the same name. His recent books are *Soundscapes of Well-being in Popular Music* (with Gavin Andrews and Paul Kingsbury) (Ashgate, 2014) and *The Afterlives of the Psychiatric Asylum* (with Graham Moon and Alun Joseph) (Ashgate, 2015).

Sarah A. Lovell is a Senior Lecturer at the University of Canterbury in New Zealand. Her background as a health geographer was shaped by undertaking her PhD in geography at Queen's University (Ontario) and postdoctoral fellowships at Queen's and Auckland Universities. Her research interests focus on structural responses to inequalities in health and health care. In recent years, this has included looking at issues of access to sexual and reproductive health care, the public health workforce and the importance of community capacity building to vulnerable communities. She has published in journals such as *Social Science & Medicine* and *Critical Public Health.*

Isaac Luginaah is a Full Professor and a Canada Research Chair in Health Geography at the University of Western Ontario, London, Canada. He is also a member of the Royal Society of Canada College of New Scholars. His broad area of research interest involves understanding of the broad determinants of population health and research on environment and health linkages. Dr. Luginaah was honored as a Paul Harris Fellow by Rotary International in 2011. In 2008, he was recognized by the Canadian Association of Geographers with the Julian M. Szeicz Award for Early Career Achievement, and in 2009 he was also recognized by the University of Western Ontario with a Faculty Scholar Award. He has been editor of *African Geographical Review.*

Sarah A. Mason is a PhD candidate and Vanier Scholar in Geography (Environment and Sustainability) at the University of Western Ontario. Sarah's current doctoral research, under the supervision of Dr. Isaac Luginaah, examines health and environmental concerns and the social and emotional impacts surrounding a regional biosolid waste to fertilizer processing facility. Sarah's broad research interests include environment and health issues, changing rural community dynamics and the impacts residents' place attachments and emotional geographies have on their well-being.

Christine Milligan is Professor of Health and Social geography and Director of the Lancaster University Centre for Ageing Research in the UK. Christine has researched and published widely in international refereed journals around the spatial aspects of ageing, care of older people and the role of the voluntary sector in the delivery of health and welfare. She has written two single authored books and four edited books on these topics and is an editor of the *International Journal of Health and Place*. Christine is expert in the use of innovative qualitative methodologies, especially those using ethnographic, participatory and narrative approaches.

Vanessa Sloan Morgan is a white settler scholar from Coast Salish territories and critical human geographer. Her research interests concern Indigenous-settler relations, with a particular focus upon sociolegal constructions of settler relationships to land and anti-authoritarian geographies. An interest in visual and arts-based methods for critical and transformative engagement has drawn Vanessa to digital storytelling: a tool that she has used in various research partnerships and educational capacities in academic and outreach settings. Vanessa holds degrees from the University of Victoria (BA, geography and anthropology), Dalhousie University (MES) and is currently a PhD candidate at Queen's University at Kingston.

Pamela Moss is Professor in Human and Social Development, University of Victoria, Canada. Her research coalesces around themes of bodies, power and knowledge production in numerous contexts. Her work draws on disability studies, social and cultural geography, feminist methodology and autobiographical writing. She uses feminism and poststructural thinking to develop analyses of topics as diverse as women's chronically ill bodies via chronic fatigue syndrome, combat soldiers' intense emotional distress via shell shock and PTSD, the making of contested illness and feminist theories of embodiment. Her most recent book, with Michael J. Prince, is *Weary Warriors* (Berghahn, 2014).

Chantelle A. M. Richmond is an Anishinabe scholar from *Biigtigong Nishnaabeg* (Pic River First Nation). Her research is framed by community-based approaches that aim to understand how Indigenous heath is affected by processes of environmental dispossession. Through this work, she seeks to aid in the preservation of Indigenous knowledge and the implementation of strategies that may foster environmental repossession efforts in affected communities. Chantelle is an Associate Professor at the University of Western Ontario, where she is Director of the Indigenous Health Lab, and Co-Director of the Indigenous Health and Well-Being Initiative. Chantelle co-wrote and produced a 60-minute documentary *Gifts from the Elders*, which focused on the preservation of Indigenous knowledge among Anishinabe communities on the north shore of Lake Superior. Her research is supported by a CIHR New Investigator Award and an Ontario Early Researcher Award. Chantelle lives in London, Ontario, with her husband, Ian, and their children, Maya and William.

Mark W. Rosenberg is a Professor of Geography and cross-appointed as a Professor in the Department of Public Health Sciences at Queen's University in Kingston, Ontario, Canada. He is the Canada Research Chair in Development Studies. In 2012, Professor Rosenberg received a Queen Elizabeth Diamond Jubilee Medal for his research on Canada's aging population, contributions to gerontology and service to older Aboriginal and non-Aboriginal populations. His research covers a wide range of topics, including health and the environment, access to health care services and vulnerable populations, especially older populations and Aboriginal peoples.

Chad Walker is a PhD candidate in geography (environment and sustainability) at Western University and a holds a Social Sciences and Humanities Research Council of Canada Fellowship. Chad earned a BA in environmental policy and analysis (Bowling Green State University 2010) and an MA in geography (University of Western Ontario 2012). His general research interests lie at the intersection of environmental science and policy – with particular reference to technologies and strategies designed to address anthropogenic climate change. With Dr. Jamie Baxter as a supervisor, Chad's current research investigates the policy instruments and local-level impacts of wind energy development across Ontario and Nova Scotia, Canada.

Rebecca Whitmore is currently completing her MA in health geography at Simon Fraser University (BC, Canada). Her thesis research is examining the retrospective accounts of Canadian medical tourists' informal caregivers. She has received a number of prestigious awards and distinctions, including a master's fellowship from the Canadian Institutes of Health Research.

Allison Williams is trained as a social/health geographer, with research interests in sense of place, therapeutic landscapes, informal caregiving and quality of life. She currently is a Professor at McMaster University (Hamilton, Ontario) and has held previous academic appointments at the University of Saskatchewan (Saskatoon, Saskatchewan) and Brock University (St. Catharines, Ontario). She currently holds a CIHR Chair in Gender, Work and Health, with a focus on caregiver-friendly workplace policies (2014–2019).

Acknowledgements

There are several people who have made this book possible: the contributors to this volume, the people at Routledge, and our esteemed editorial assistant. We thank all of the contributors. Personalized accounts of researching, and reflections on praxis, can be a lot to ask of a researcher, and we appreciate the frankness and honesty. Their work has pried open the spaces in between the 'doing of research' and, in the process, has led us to critically ponder on how research unfolds – thus, expanding health geography practice.

We are grateful for the assistance, patience and indulgence of the people at Routledge ushering the book along to completion: Katy Crossan, Carolyn Court and Will Tyler in particular. Thanks go to Susan Elliott and Allison Williams too in their roles as series editors for the Geographies of Health Series.

We would particularly like to thank Leah Graystone for her dedication and commitment in providing crucial and timely editorial assistance that made all things work smoothly and kept us on track.

Preface

The practice of qualitative research

While driving to interview 'Barbara' I listen to Chad's interview with her from a few months before. She has provided a more or less steady stream of emails about the fight against the turbines, and this specific visit is about the initial interpretations of Chad's interviews with 23 residents. I have talked to Barbara and others on the phone and over email – they have a group opposed to the turbines. The group is accused by others of being selfish, being anti-environment, and is ridiculed for claiming their ill health could be caused by turbines. This is why those in the group are adamant that we get it right – the turbines must go. I am guarded; they want something very specific from Chad and I – something we did not promise. A document summarizing the preliminary findings prompted written insults from them about the integrity of our research – betrayal. The interview audio does not finish by the time I arrive, but it is clear Barbara is hurting – heart-wrenching. What have I gotten Chad into?

I meet Barbara in somebody else's home – she cannot sleep at her own place because of the turbines – a very small turbine spins nearby, presumably helping to power the building, an eco-vehicle in the driveway – 'anti-environment' indeed. She is welcoming, kind; we have a long talk; there are tears – crocodile alarms stay mute. We later visit her unsellable house; it is not clear where home is. The turbines are close, so is the lake – beautiful vista – we chat about how all of this looks on a map. I am invited to snap photos around the property, including some of placards from a recent protest – easily recognized as hers. Chad and I have several long chats – we have bit firm on the hook and line of clean green wind energy. Everyone ridicules people like Barbara though; let's not participate. Barbara has told her story in the media, as have others with shared experiences – the Premier decries NIMBY. This goes beyond 'representing' my morning with Barbara. Our publications are mixed methods and include a dash of quotes from Barbara's original interview amid survey findings, a map and some statistics showing that turbines cluster too tightly in Barbara's neighborhood. Barbara's story, my morning with her, now part of the flotsam and jetsam of what we think people will listen to. The politics of praxis. There are other paths though, and this is why it was easy to say 'yes' when Nancy approached me about putting together this book.

Jamie Baxter, London, ON

A breach . . . 'Have you ever injected an Epi-pen?' Kyle asked me. I thought, *What now? No, I've never used one.* 'Have you seen the needle before?' *No, not the needle.* I felt nervous about losing control of the interview, but kept responding anyways. 'I like to call it my lifeline because it's your last resort; you never want to use it. If you are on the ground and can't breathe, that is your life.'

'Give me like two seconds' – I watched as Kyle stood up, left the room, and re-emerged with a black case and an orange. He put both in front of me and directed me to open the case. I pulled out the needle. I was shaking, but forced myself to appear calm. I followed his instructions without question – 'Pull off this gray tip, push down on the orange and hold for 10 seconds. Pretend this orange is your leg. Screw the top off, pull it right up, that's the needle, pop! . . . It's not huge, but that is your life in that brief moment.'

Kyle was 16 years old with life threatening food allergies and afraid of needles. He took over the interview – he pushed me into his day-to-day experiences of risk to get a glimpse of his fears. Years later, as I reflect upon Kyle's interview, I realize the experience ruptured my firmly held assumptions of doing research, my role as researcher, and what I had previously thought of as tidy, consistent methodology. Kyle changed how I explore, interpret and represent the fullness of a participant experience and gave me permission to drag onto the surface the uncertainties, conflicts, fears and confusions that research 'in the field' can bring. The chance to pull together a qualitative research collection that makes transparent the fullness and diversity of 'doing of research' was an opportunity too good to miss.

Nancy E. Fenton, Hamilton, ON

Our purpose in this book is to make visible the fullness and diversity of doing research – the challenges in undertaking research and the connections between knowing and doing the practice of qualitative research. We have assembled a collection of chapters that showcase how the contributors confront the uncertainties and ambiguities of practice and, through reflection on praxis, illustrate how knowledge generation is a complex process of interrelated interpretations. This collection of essays tells the fullness of qualitative research stories 'in the field' that at times, is unsettling – filled with false starts, impasses and surprises that inevitably make visible the cost that the dilemmas carry for researchers. Too often, dilemmas and moral choices are tucked away from public scrutiny in the name of 'doing good research'. In this collection, we ask authors to make these challenges transparent in order to open the crevices of policy and practice, which cultivates the in-between spaces of our work.

In *Practicing Qualitative Methods in Health Geographies*, we showcase health geographers who are designing theoretically informed research agendas and who are turning to a diverse range of interpretative methodologies for exploring the complexities of health, illness, space and place to gain more comprehensive understandings of well-being and broader social models of health and health care. These contributors join a growing number of qualitative researchers who are drawing upon postmodernism to examine the ontology of geographic discourse and who are concerned with issues of representation, the body and health-care policy. In this volume, authors employ a host of creative qualitative methods to explore the psychosocial experiences of individuals more directly, using such traditional methods as in-depth interviews and group discussions, participant observation, diaries and discourse analysis, but also more novel techniques such as 'go-along interviews', reflexive writing, illustrations and photographic techniques.

We hope this volume will appeal broadly to health geographers and will especially serve to strengthen novice researchers' confidence to adapt and refine existing theories and methods in their own research, and to reflect on their practice through the interests and values served by their projects. The aim of the book is not to provide a generic methodological 'how-to guide' that can be deployed – a tool kit or guide does not exist to tackle the complexity of our work simply. Rather, we suggest a 'living practice' – more complex, nuanced and ambiguous rather than 'technical skill and competence' (Higgs, et al., 2011). We reflect on praxis that embraces a need to creatively open up intellectual spaces to explore and articulate our practice.

The creativity of the contributors in this volume comes from seeing things in new ways and expressing it fully to open up our thinking and our insights. The authors draw upon expressive, arts-based and non-representational methods to capture the fullness of what is happening in the social world and to re-present their research findings in new ways. This attentiveness to the nuances and dimensions in ordinary life has the capacity to expand our seeing, thinking, feeling and being in the multilayered and ambiguous space where power and inequities reside. We believe such creative work has application across a broad spectrum of health geography; it has the potential to richly describe practice phenomena in such a way that we can appreciate, come to know and more deeply understand the social constructions of place, space, experience and meaning.

References

Higgs, J., Titchen, A., Horsfall, D., and Bridges, D. (eds.), 2011. *Creative Spaces for Qualitative Researching: Living Research.* Rotterdam: Sense Publishers.

Abbreviations

APIs	Application programming interfaces
CAQDAS	Computer assisted qualitative data analysis software
CBPR	Community-based participatory research
CBR	Community-based research
CIHR	Canadian Institutions of Health Research
EAGLE	Effects on Aboriginals for the Great Lakes environment
FCG	Family caregiver
GIS	Geographic information system
IK	Indigenous knowledge
KT	Knowledge translation
LAC	Local Advisory Committee
LGBTTQTS+	Lesbian, gay, bisexual, transvestite, transsexual, queer, two spirit
LOI/CF	Letter of information/consent form
ME	Myalgic encephalomyelitis
NHS	National health service
NRT	Non-representational theory
O/S/P/V	Objects, spaces, possessions, or views
P/EOL	Palliative and end-of-life
PPE	Participatory photo elicitation
REB	Research ethics board(s)
TCPS 2	Tri-Council policy statement 2

1 Praxis in qualitative health geography

Jamie Baxter and Nancy E. Fenton

Introduction

This book explores relatively novel qualitative methodologies from the point of view of sharing experiences of doing research. It is not meant to teach specific research methods, not exactly at least; instead it is meant to show what happens when theory and method meet in the real world. This book is less about the how-to of qualitative research and more about how-to-learn-from qualitative research in action from researchers who have been asked to critically reflect on their research journeys. To nudge the issue somewhat, each chapter begins with a section entitled 'Reflections on Praxis' where the authors are asked to critically reflect upon and write about the special challenges they faced conducting their particular form of qualitative research.

Health geographers are increasingly designing theoretically informed research agendas and are turning to a diverse range of interpretative methodologies for exploring the complexities of health, illness, space and place to gain more comprehensive understandings of well-being according to broader social models of health and health care. While traditional approaches to qualitative research – ones that employ such methods as interviews, focus groups and ethnography for gaining depth and nuance of understanding – may remain the 'bread and butter' of health geography, a growing number of health geographers are pushing the boundaries of qualitative methodology to interrogate the epistemology, ontology and ethics of geographic discourse. Geographers who are being innovative with qualitative approaches are concerned with a range of issues such as representation, the body, environment and informal social arrangements in their use of modified methodologies. These methodologies include: go-along interviews, community-based participatory designs, reflexive writing, narratives solicited without interviews and photographic techniques to represent experiences of health, illness and health care, while others have taken up recent calls within geography to go beyond representation itself. As Milligan points out in her chapter, and cites Thrift (2000), qualitative researchers need to be 'more methodologically imaginative', but at the same time share the experiences of what such imaginings have wrought by reflecting publicly on praxis.

Being methodologically imaginative means taking seriously that everyday life will contain a number of dilemmas because it comprises of contrary themes so that each person will never confront the world with a stable and unified identity

(Billig, 1991; 1996). This is particularly the case, we would argue, when we face dilemmas in our research practice. At this point in research practice, decisions have to be made and an opportunity has to present itself for a reflexive standpoint to emerge about wider structural processes driving and influencing our data collection. As Becker (1965, p. 602) reminds us, 'no matter how carefully one plans in advance, the research is designed in the course of its execution. The finished monograph is the result of hundreds of decisions, large and small, whilst the research is underway' (Roberts and Sanders, 2005).

Defining praxis

We view praxis and practice as closely aligned in the context of this book. Praxis is meant as a signal that the book concerns writing about the *process* of research and reflection and not simply the practical nuts and bolts of deploying a method – the latter being what might typically be regarded as research *practice*. However, what we mean by *praxis* in this volume may not be what critical geographers in particular may anticipate. For some praxis is about taking a more activist stance as academics, particularly in the context of critical and radical (e.g., Marxist) social theory that aims at positive social change through research and writing (Fuller and Kitchen, 2004). Praxis defined as putting theory into action through research has thus been focused on social goals like gross income, health inequalities and other forms of health, social and environmental injustice. Though that definition does apply here too, we are also using praxis in a broader sense; for example, as Parr (2004, p. 249) defines it in terms of 'activism (direct action) that potentially might make a difference to inequitable processes'. While the authors in this volume would likely agree with the goal of 'making a difference', they may not self-identify as activists taking direct action through their research – though some certainly do take on such a direct action stance (e.g., Lovell and Rosenberg, Chapter 6; Richmond, Chapter 9).

This book is focused less though on the activism aspect of praxis and more on the process of reflexivity – exposing the warts and all of qualitative research 'in the field'. While qualitative researchers are routinely reflexive when doing research, their experiences, conundrums and even missteps exposed through reflexivity are not routinely written down to be shared with other researchers. In this sense, praxis, as is defined here, falls between the cracks of writing about theory, writing about methodology, writing in the field (i.e., field notes) and writing about research implications (i.e., knowledge mobilization). Praxis focuses on the spaces in between the 'doing of research' to critically ponder on how the research unfolded – revealing the elements of empirical engagement with participants that went awry and trumpeting the features that led to better-than-anticipated responses and outcomes.

Theory as motivation for change in qualitative praxis

This movement towards more creative qualitative geographies has been encouraged by ongoing calls for greater attention to social theory in health geography

(Litva and Eyles, 1995), including Kearns (1993) who urges anchoring health geographic work to the core concept of place rather than dwelling entirely on biomedical notions of disease (Fleuret and Atkinson, 2007). Thus, socially, culturally and spiritually, contingent illness takes center stage as people are increasingly conceptualized less as vessels of disease and more as people who experience physical health challenges in everyday settings that may improve or worsen well-being. Though formal health care systems remain a focus, so too are informal systems of coping and care, which are embedded in social and cultural processes linked to family, work and community. One of the landmark events of the time was the launch of the journal *Health and Place* in 1995 that created a specific outlet for qualitative researchers interested in meaning-oriented research and methodologies focused on such principles as proximity, nuance, depth, power, critique and change.

For several decades now the concept of *place* defined as relational, meaning-laden space – something that is contextual and socially dynamic – has opened doors for qualitative researchers in the same way that *space* – something physical containing compositional factors as variables (e.g., race, gender, social class) – has done for quantitative researchers. Methodology has also evolved, drawing more tentative lines between these two broad categories of approaches to research in such form as 'questerviews' (Adamson, et al., 2004). Others prefer to embrace the integration of quantitative and qualitative methods in the same study by practicing them separately to more fully understand health and its determinants. For example, the wealth of health inequalities research continues to use a variety of largely quantitative methods to tease out how place as neighborhood ('context') and individual-level factors ('composition') influence health (Cummins, et al., 2007; Macintyre, et al., 2002; Smith and Easterlow, 2005); while Bolam, et al. (2004), picking up on themes developed by Eyles and Donovan (1986) decades before, predominantly use interviews to go deeper and explore how class identity and its impacts on health are reproduced and resisted. To better understand the contingencies of place and health, others have embarked on methodologies as a way to bring us even closer to these experiences in neighborhoods through such methods as go-along interviews (Carpiano, 2009).

Qualitative research, oriented within more place-focused health geography was sparked by the 'cultural turn' in the wider discipline and the social sciences. Yet, critiques from critical theorists and others have helped to push forward both theory and methodology. The 'turn', starting as early as the 1970s, was a shift away from positivism and its limitations (e.g., reductionism) towards social construction and its understandings of the social world within groups in terms of cultures and meanings. In health geography those groups have been defined in numerous ways; for example, health professions (e.g., Marshall and Phillips, 1999), women (e.g., Garvin and Wilson, 1999) and Indigenous peoples (Castelden and Garvin, 2008; Richmond, et al., 2007). Though the cultural turn has come under attack in the broader discipline of geography, particularly from critical theorists, Parr (2004) suggests that health geographers have insulated themselves somewhat from such debates. Instead, as a group we seem willing to welcome cultural and critical-theory approaches equally – despite any ontological and epistemological incompatibilities they might ostensibly entail. That said, Parr argues there is a danger of culturally oriented research becoming 'too subjective', rendering it unhelpful for meaningful policy change and believes more attention needs to be paid to emancipatory politics in health geographic research.

Qualitative health geography researchers have also adapted, merged and shaped cultural and critical approaches to focus on the body, environment and landscape. For example, Hall (2000) uses concepts like 'embodiment' to articulate how health geographers may move beyond the body as simply a biological entity 'invaded by disease' to something that is *represented* in various ways that is inscribed with identity, culture and even politics. Greater attention has also focused on affect and emotion, which tend to involve deep, detailed and conceptually rich ethnographies (Pile, 2010). Such approaches further encourage qualitative health geographers to explore potentially new insights in arcas such as mobilities and mental health research along with feminist studies (see also Moss and Dyck, 2003). Meanwhile, environment and health geographic research focuses attention on environmental hazards as threats both to bodies and communities. Qualitative methods have been used to explore the social construction of meaning and conflict over such health-harming hazards as fires (Eyles, et al., 1993), waste sites (Elliott, et al., 1993), petrochemicals (Atari, et al., 2011) and pesticides (Hirsch, et al., 2010). Likewise, therapeutic landscapes research focuses on the relation of bodies to physical environments, but with less emphasis on health threats and more emphasis on the restorative relationships people have within particular settings (Gesler, 1992; Williams, 2007; Wilson, 2003). For instance, using interviews and media analysis, Wakefield and McMullan (2005) argue that landscapes may simultaneously hurt and heal depending on the physical and social locations of the people involved.

Political ecology is another theoretical thread within the social sciences that has been taken up by geographers interested in both disease and health as wellbeing. Through integrating multiple methods, political ecology has played a part in broadening the scope of qualitative methodologies. Political ecological work focuses on the relationship between social, political and environmental systems and their links to health. These particular studies tend to take on a critical theoretic flavor by providing alternative readings of hegemonic discourses on health – for example, sociopolitical and discursive dimensions of disease, the social production of nature and the intimate relationship between socially reproduced nature and health. For instance, King (2009) explores how these phenomena can shape how disease is managed, using case examples of HIV/AIDS in South Africa and cholera in Zimbabwe.

These various theoretical perspectives and concepts in one way or another at least implicitly, and more often explicitly, advocate for the use of qualitative methods in health geography. This has prompted more thorough consideration of why and how we deploy qualitative methodologies. It is in this context that health geographers have contributed to methodological debates that extend well beyond health geography. In *The Professional Geographer*, a focus section on 'qualitative approaches in health geography' detailed some of the early issues facing qualitative researchers, yet unwittingly at the same time brought into focus the dominance of face-to-face, in-depth interviewing among qualitative researchers (Elliott, 1999). In that issue, Wilton (1999) and Baxter and Eyles (1999) explored the benefits of using in-depth interviews for understanding the meaning of HIV/AIDS and environmental health risk respectively in everyday life. Robinson and Elliott (1999) used interviews in a critical analytic mode to understand informal systems of community-based heart health promotion. In contrast, Marshall and Phillips (1999) interviewed physicians to understand professional referral practices in the formal health care system in the UK. In that same issue though, Garvin and Wilson (1999) push us to consider more creative uses of interviews and narrative analysis by adapting them to incorporate other media – such as photos or other images. They creatively employ photos of tanned and untanned bodies to explore issues of body image and sun safety among women, by asking them to tell stories about the women in the photos.

Pushing beyond the traditional marriage between interviews and humanistic geography, Dyck (1999) and Cutchin (1999) advocate for more theoretically informed methodological choices through more thorough engagement with the relationship between theory and method. On one hand, Dyck builds upon themes previously developed by Kearns (1993) highlighting the opportunities for using qualitative methodologies to deconstruct conceptual categories (e.g., space, disease, cause, care) and to change power imbalances and inequities (i.e., feminist approaches) central within medicine and traditional medical geography. On the other hand, Cutchin (1999) focuses on the philosophical bases of qualitative research in health geography, urging us to consider pragmatism as a foundation as well. Indeed pragmatism is consonant with the social constructionism that is often implicit in much of health geographic work, yet Cutchin articulates a broader understanding of human experiences and human actions – both core constituents of place. Thus, by adapting theories from geography and the social sciences into a health geographic frame, health geographers have had cause to rethink their use of method within theoretically informed methodologies – issues that are worth teasing apart further.

Method, methodology and ethics

As the title suggests, this book is ostensibly about 'methods', but it is also about methodology with some blurring of the distinction throughout this volume. Method is commonly referred to as a tool for accumulating data to be analyzed (e.g., interviews) while methodology (e.g., community-based participatory research) refers to the combination of those tools framed within a theory or concepts that guide how and why data are to be collected. Often theory makes reference to how the world is organized (i.e., ontology) and sets out principles for what counts as relevant data (i.e., epistemology) and assumes moral and ethical values (e.g., working against marginalization towards justice). The blurring between method and methodology also includes analysis, particularly in the sense that in qualitative research, analysis occurs throughout the research process. Thus, there may be reference to 'analytical methods' (e.g., ethnography) whereby data are interpreted more fluidly as they are collected. In this sense, conceptual categories may continually be reshaped, abandoned or resurrected as the research process progresses, rather than amassed as a systematically gathered corpus of material (e.g., documents, interviews) *before* analysis begins.

The blurred lines between theory, method and methodology are further symbolized by the advent of non-representational theory (NRT) in geography (see Chapter 12). In simple terms, NRT focuses on how life takes shape and gains expression in the social world. It makes reference to how we sense and know our world to better define alternative ways of apprehending it – alternatives that go beyond speech and text to understand events before they are given meaning through social construction (e.g., in an interview). Though to date, there is admittedly more emphasis on theory and methodology than there is on how exactly this apprehending might happen (see Hanlon, 2014; Kearns, 2014). Choosing to employ 'theory' or 'analysis' (e.g., discourse analysis) or 'methodology' (e.g., feminist methodology), or 'method' (e.g., photo elicitation method) as sublabels is in this sense, tends to signal where emphasis lies. However, in most cases the NRT label is meant to convey something that integrates theory, method, analysis and methodology together into a coherent whole.

The distinction between a method and methodology is not always clear as few qualitative researchers deploy methods without probing the philosophical and theoretical underpinnings of their research choices. For example, Harding (1989, p. 18) addresses the debates about the existence of a feminist method, concluding that the questions are 'misguided and should be abandoned' because feminists do what they do by paying attention to gendered power relations regardless of the tools they use to collect and analyze data. In this regard, moral and ethical imperatives relating to justice and empowerment may supersede method, acknowledging of course that there is a wide variety of methods which may serve to achieve these ends and perhaps even feminist 'ways of knowing and doing' that need to be taken into account (Moss, 1993, p. 48).

Many of the contributions in this book point to the idea that the *ethics* of method and methodology are similarly slippery. Ethics in health geography implicitly and explicitly thread through the way stories are structured and how narratives are silenced, contested or accepted. For instance, both Mason et al. (Chapter 4) and Williams (Chapter 13) in this volume write directly about the paradoxical and sometimes dubious role of university research ethics boards (REBs), particularly the increasing emphasis on rule-bound procedures for satisfying 'ethics' (Dyer and Demeritt, 2009) – in other words, 'ethics creep' (Haggerty, 2004). These two chapters illustrate several ways in which complex and legalistic procedures for achieving informed consent may be in tension with or are outright antithetical to what may be viewed as the professional ethics of being true to a theory and praxis of socially relevant and transformative social science. Though the tensions between REB ethics and professional ethics is far from resolved in the chapters of this book, the reflections on praxis do serve to highlight the need for greater sensitivity and awareness of such issues in hopes that lines may start to be drawn in the sand to redress problems of increasingly rule-bound and prescriptive systems which dictate how we operate in the field.

It is an exciting time to be practicing qualitative research in health geography because there is a wealth and diversity of ways to tackle the nuanced dimensions of research problems within the field. The complex landscapes of health geography contain a wealth of lived experiences that connect webs of relationships where no aspect can be fully understood without considering the whole. Different ways of knowing and of relating are essential to explore these experiences. Health geographers engaged in research understand that health and care are multilayered, holistic and ambiguous, and that research often reveals interpretative uncertainties – a dimension of depth where hierarchies of power and inequalities are more evident. More and more, health geographers are creatively deploying qualitative methods and new paradigms of inquiry to the questions that arise from within these webs to greatly enhance our understandings of the social constructions of place, space, experience and meaning.

Organization of the book

Each chapter begins with a 'Reflection on Praxis'. The authors were specifically asked to reflect on the special challenges they have faced conducting the empirical research they describe and to expose aspects of that work that might not otherwise fit neatly into the chapter, including such topics as philosophical/moral dilemmas, institutional or disciplinary resistance, technical challenges and issues that needed to be resolved along the way and would be instructive to others considering the methodology described. This is a rather large ask of our contributors, but it is essential for molding and developing these methodologies. We anticipate that the lure of these relatively innovative approaches will prove irresistible for those who have a sense that traditional ways of doing things, though still relevant and useful, only take us so far.

The contributions begin with a three-chapter section on ethics and power in qualitative health research. Christine Milligan introduces narrative correspondence method in Chapter 2. This method asks participants to write a narrative of their experience without the aid or presence of the researcher to help co-construct the narrative. Milligan argues that such an approach is intended to break down power imbalances by empowering the participant to take the lead on developing a narrative with only initial framing by the researcher. Like Williams, Milligan is interested in qualitatively exploring informal carers – those caring for older people in New Zealand. The method and empirical work described in the chapter highlight how place, care and identity are interconnected over time and space. Milligan illuminates the idea that by telling stories of illness the 'changed body once again becomes familiar' to the storyteller and allows others to bear 'witness to conditions that medicine cannot describe' (see also Moss, Chapter 5). Details of recruitment strategies are elaborated upon, and Milligan recommends a fourfold categorization of narratives to make sense of the resultant stories: ontological narratives (stories actors use to make sense of their own lives), public narratives (stories used to make sense of events and institutions), metanarratives (stories that embed actors in wider society) and conceptual narratives (the narratives researchers use to describe the other three forms of narrative).

In Chapter 3, Tara Coleman traces a go-along approach of sorts by sharing in the experience of creating photographs while interviewing research participants in their homes about aging in place. Thus, her presence and participation in taking the photos distinguishes her approach from other forms of photo elicitation like photovoice, which encourages participants to take photos on their own (Castleden and Garvin, 2008; McIntyre, 2003). In this chapter, Coleman emphasizes that the photo elicitation method frees up the participants to focus less on technical details of running equipment and more on the composition and content of the image. This co-construction of the photo is meant to give over control of the photo in these in-depth encounters while at the same time to help break down social barriers and build rapport – indeed intimacy. Coleman uses Polly's story (complete with photos) to illustrate the rich depth of understanding that can be achieved using this methodology. Along the way she gives a sense of some of the challenges of conducting this type of research along with the potential impact such research can have on policy change.

In Chapter 4, Sarah A. Mason, Chad Walker, Jamie Baxter and Isaac Luginaah describe how the politics of fieldwork in the context of environment and health controversies can result in a backlash from activists. What happens when the people in the community you are trying to help try to shut down your study? They problematize such issues as taking sides and empowerment in community-based research since environment and health disputes can include very savvy and resourceful publics who are nevertheless marginalized by questionable environment and health policies. Using concepts from ethics and qualitative research, the authors explore experiences in two separate studies and expose the need to reframe discussion and debate about such phenomena as insider status, member checking and professional versus research-board ethics.

The second section, which concerns representation, self and community begins with a chapter involving autobiography. Pamela Moss in Chapter 5 employs autobiographical analysis using a feminist approach to create a richly detailed and personal account of the experience of illness in everyday life – a critical approach that may be used to contest prevailing discourses in health policy. Such research serves to blend several areas of inquiry of interest to health geographers, including bodies, power, space and feminism. The goal of this methodology, she writes, is to 'humanize disease', to better inform health policy as it relates to traditional spaces within the formal health care system, but also to think about other spaces like home and work as sites where positive efforts can support improved well-being. Moss provides windows into her own illness autobiography as a way to explore some of the methodological and substantive debates and challenges involved in such work: bracketing, illness identities, relational notions of experience, discourse, materiality, enactment and perhaps most importantly challenging taken-for-granted discourses of illness and coping.

Sarah A. Lovell and Mark W. Rosenberg also write about community-based research in Chapter 6, focusing on community capacity building through participatory qualitative research. They explore whether and how building community capacity – the skills and abilities of a group to address their own health problems – is an achievable and measurable research outcome. Like the Moss chapter, they draw heavily on feminist methodology and the practice of reflexivity as key strategies to attend to issues of power and exploitation in participatory justice-oriented methodologies. Examples drawn from an accessible transit project supporting people with disabilities highlight some of the challenges of transformative research and takes seriously the ethics of more thoroughly engaged research. They explore a range of issues, such as flexibility, responsiveness, attrition and sustaining momentum. Through their experiences Lovell and Rosenberg challenge researchers to think further about how to match researcher and marginalized community goals for improving the daily lives of the latter.

In Chapter 7, Jennifer Dean introduces the use of go-along interviews within qualitative methodology. She draws upon her experiences studying adolescent body weight, healthy eating and physical activity with a particular focus on youth in low socioeconomic status neighborhoods. Her experiences in the field provide unique insights into the challenges of using this methodology as well as issues navigating research ethics boards. She emphasizes the advantages of an approach that allows for unique interactions between researcher and research participants within the participants' environments; an approach that avoids environmentally deterministic interpretations of urban design meant to increase neighborhood mobility. Dean deftly situates go-along interviews within three theoretical and methodological trends in health geography: (1) the new mobilities paradigm, (2) non-representational theory (see Chapter 12), and (3) relational theory. That is, these all encourage thinking beyond non-traditional modes of enquiry, ones that move us away from understanding the meaning of places through utterances and text alone.

As the opening chapter to the section on visual media, Stephanie E. Coen challenges us to reflect on the fact that we often 'think through writing', an issue she explores in Chapter 8 on the use of drawing as a way to elicit meaning about health-related activities. The chapter includes a review of the literature on participant-generated drawing as a method for exploring health geographies. Among other things, allowing participants to explore their thoughts, meanings and emotions through drawing shifts the balance of power in the research relationship by casting image authors in the role of experts. In the latter half of the chapter, Coen reports on a pilot study involving drawings from university gym users along with textual descriptions and responses to guided questions. The pilot study explores the possibilities of using participant-generated drawing in a distanced way, and Coen explores the advantages of conducting similar work in a face-to-face interview context as well.

In Chapter 9 Chantelle A. M. Richmond provides a unique research project built upon a community-based participatory research (CBPR) methodology that combines individual interviews, group interviews, observation and training that culminated in a video for the community. The research goals were both practical and academic – bringing First Nation (Anishinabe) Elders together with youth from the community to better understand views on health through honoring the linkages between health, environment, history and culture. The research team played a central role in organizing the work of training youth on how to conduct in-depth interviews – skills they later employed to interview Elders in their communities, many of which were video recorded. Graduate students played important roles that were carefully woven into the research design to acknowledge ethical issues of power and trust. Richmond points out that although there is considerable theory and writing about the reasons for conducting CBPR, hers is one of a few empirical *examples* of how they play out in practice – including lessons learned.

Heather Castleden, Vanessa Sloan Morgan and Aaron Franks build on traditional photovoice to engage Indigenous community members in digital storytelling in Chapter 10. Like traditional photovoice the participant is encouraged to produce or gather imagery that is ultimately transformed into a video story. Castleden and colleagues recount how they have used such techniques within a broader CBPR methodology. Unlike Richmond's chapter, which also engages a CBPR and the use of more traditional documentary style video, the digital storytelling allows the storyteller to take on the role of narrator – the person behind, rather than in front of, the camera. This speaks to issues of control in the research process. The chapter provides strategies for using a workshop format to manage the process; to separate the creation of the stories from the technically challenging issues of using the software and hardware needed to produce the videos.

S. Michelle Driedger and Theresa Garvin provide strategies for analyzing news media in health geography in Chapter 11. They make a key distinction between the challenges of analyzing traditional media based on a sender-receiver model and modern forms of media which allow for more participation and shaping of messages by various publics. At the same time they point out ways large media outlets continue to control those same messages in various ways online. For example, they highlight that online news media which allows interactivity through commenting may breech anonymity and confidentiality assumptions of traditional media data and thus require research ethics clearance. Further they highlight the paradox of using seemingly broad and relevant keyword search filters in public news databases that may actually exclude relevant news items unexpectedly. These problems can be amplified when conducting internet search-engine based searches of social media, blogs and other forms of information sharing and networking. Driedger and Garvin set all of this in the context of using such secondary data to qualitatively analyze health geographic topics in a way that balances depth, creativity and accuracy with the actual resources available to conduct rigorous media analysis.

The final section is comprised of three chapters on (non)representation, affect and social life. Gavin J. Andrews and Eric Drass in Chapter 12 introduce non-representational theory using two case examples involving musical pieces. Non-representational theory has recently attracted much discussion by geographers to the point that Andrews and Drass suggest one would have to have been 'living on an island as a qualitative human geographer' not to have in one way or another engaged with non-presentational theory. This growth in interest is attributed to an increasing dissatisfaction with the capacity for in-depth interview methods, commonly used in humanistic geography, to capture the fullness of what is *really* going on in the social world. The authors encourage us to stretch our thinking about how the various senses may be engaged in research and (non)representation by challenging us to listen and see more broadly, indeed 'pre-cognitively'. Such work has potential application across a broad spectrum of health geography including healthy body imagery, aging, disability, mobility and therapeutic landscapes.

Allison Williams discusses the challenges in conducting empirical research with relatively vulnerable groups who are managing health issues in Chapter 13. She recounts experiences and lessons learned through studying the impact of health care reform on informal family caregivers and discusses: (1) ethical issues, (2) practical issues of accessing and recruiting participants, and (3) emotion work in palliative care research. By sorting through these issues, Williams highlights the research conundrum that the people most in need of policy change are often also among the most vulnerable populations to study. For anyone interested in studying such populations directly, this chapter highlights important lessons for planning and conducting such research, magnifying benefits and minimizing the potential for harm to the research participants and researchers themselves.

Like Milligan, Valorie A. Crooks, Victoria Casey and Rebecca Whitmore explore informal caregiving in Chapter 14, by describing challenges of conducting secondary analysis of in-depth interviews. The authors suggest the possibilities for asking new questions of old data; in this case, data that were gathered for other purposes, to answer different questions. The authors provide an example of analyzing telephone interviews with Canadian medical tourists which were originally meant to better understand the experiences of the interviewees themselves; but are explored in the chapter for understanding the experiences of their caregiver companions (e.g., spouse, friend). This chapter highlights both the challenge and limitation of such work (e.g., (mis)representation, voice), yet encourage researchers to stay attuned to the unanticipated by remaining flexible and open throughout the research process.

References

Adamson, J., Gooberman-Hill, R., Woolhead, G., and Donovan, J., 2004. 'Questerviews': Using questionnaires in qualitative interviews as a method of integrating qualitative and quantitative health service research. *Journal of Health Services Research & Policy*, 9(3), pp. 139–45.

Atari, O., Luginaah, I., and Baxter, J., 2011. 'This is the mess that we are living in': residents everyday life experiences of living in a stigmatized community, *GeoJournal*, 76(5): 483–500.

Baxter, J., and Eyles, J., 1999. The utility of in-depth interviews for studying the meaning of environmental risk. *The Professional Geographer*, 51(2), pp. 307–320.

Becker, H., 1965. *Outsiders: studies in the sociology of deviance*. New York: Free Press. pp. 41–58.

Billig, M., 1991. *Ideology and opinion: studies in rhetorical psychology*. London: Sage Publications.

Billig, M. 1996. *Arguing and thinking: a rhetorical approach to social psychology* (2nd ed.). Cambridge, UK: Cambridge University Press.

Bolam, B., Murphy, S., and Gleeson, K. (2004). Individualisation and inequalities in health: a qualitative study of class identity and health. *Social Science & Medicine*, 59(7), pp. 1355–1365.

Brown, T., and Duncan, C., 2002. Placing geographies of public health. *Area*, 34(4), pp. 361–369.

Carpiano, R., 2009. Come take a walk with me: the "go-along" interview as a novel method for studying the implications of place for health and well-bring. *Health & Place*, 15(1), pp. 263–72.

Cummins, S., Curtis, S., Diez-Roux, A., and Macintyre, S., 2007. Understanding and representing 'place' in health research: a relational approach. *Social Science & Medicine*, 65, pp. 1825–38.

Castleden, H., and Garvin, T., 2008. Modifying Photovoice for community-based participatory Indigenous research. *Social science and medicine*, 66(6), pp. 1393–1405.

Cutchin, M., 1999. Qualitative explorations in health geography: using pragmatism and related concepts as guides. *The Professional Geographer*, 51(2), pp. 265–274.

Dorn, M., and Laws, G., 1994. Social theory, body politics, and medical geography: extending Kearns's invitation. *The Professional Geographer*, 46(1), pp. 106–110.

Dyck, I., 1999. Using qualitative methods in Medical Geography: Deconstructive moments in a subdiscipline? *The Professional Geographer*, 51(2), pp. 243–53.

Dyer, S., and Demeritt, D., 2009. Un-ethical review? Why it is wrong to apply the medical model of research governance to human geography. *Progress in Human Geography*, 33(1), pp. 46–64.

Elliott, S., 1999. And the question shall determine the method. *The Professional Geographer*, 51(2), pp. 240–43.

Elliott, S., Taylor, S. M., Walter, S., Stieb, D., Frank, J., and Eyles, J., 1993. Modeling psychosocial effects of exposure to solid waste facilities, *Social Science & Medicine*, 37(6), pp. 791–804.

Eyles, J., and Donovan, J., 1986. Making sense of sickness and care: an ethnography of health in a West Midland town. *Transactions of the Institute of British Geographers*, 11(4), pp. 415–27.

Eyles, J., Taylor, S. M., Johnson, N., and Baxter, J. 1993. Worrying about waste: living close to solid waste disposal facilities in Southern Ontario, *Social Science & Medicine*, 37(6), pp. 805–812.

Fleuret, S., and Atkinson, S., 2007. Wellbeing, health and geography: a critical review and research agenda. *New Zealand Geographer*, 63(2), pp. 106–118.

Fuller, D., and Kitchin, R., 2004. Radical theory/critical praxis: academic geography beyond the academy. In *Radical theory, critical praxis: making a difference beyond the academy*. Victoria, BC: Praxis (e)Press. pp. 1–20.

Garvin, T., and Wilson, K., 1999. The use of storytelling for understanding women's desire to tan: lessons from the field. *The Professional Geographer*, 51(2), pp. 297–306.

Gesler, W. M., 1992. Therapeutic landscapes: medical issues in light of the new cultural geography. *Social science and medicine*, 34(7), 735–746.

Haggerty, K., 2004. Ethics creep: governing social science research in the name of ethics. *Qualitative Sociology*, 27(4), pp. 391–414.

Hall, E., 2000. 'Blood, brain and bones': taking the body seriously in the geography of health and impairment. *Area*, 32(1), pp. 21–29.

Hanlon, N., 2014. Doing health geography with feeling. *Social Science & Medicine* 115, pp. 144–46.

Harding, S., 1989. Is there a feminist method? In Tuana, N. (ed.) *Feminism and science*. Bloomington: Indiana University Press. pp. 18–32.

Hirsch, R., Baxter, J., and Brown, C. 2010. The importance of skillful community leaders: understanding municipal pesticide policy change in Calgary and Halifax. *Journal of Environmental Planning and Management*, 53(6), pp. 743–757.

Kearns, R., 1993. Place and Health: towards a reformed medical geography. *The Professional Geographer*, 45(2), pp. 139–147.

Kearns, R. (2014). The health in 'life's infinite doings': A response to Andrews et al. *Social Science & Medicine*, 115, pp. 147–149.

Kearns, R., and Moon, G., 2002. From medical to health geography: novelty, place and theory after a decade of change. *Progress in Human Geography*, 26(5), pp. 605–625.

King, B., 2009. Political ecologies of health. *Progress in Human Geography*, 34(1), pp. 1–18.

Litva, A., and Eyles, J., 1995. Coming out: exposing social theory in medical geography. *Health and Place*, 1(1), pp. 5–14.

Mayer, J. D., and Meade, M. S., 1994. A reformed medical geography reconsidered. *The Professional Geographer*, 46(1), pp. 103–106.

Macintyre, S., Ellaway, A., and Cummins, S., 2002. Place effects on health: how can we conceptualise, operationalise and measure them. *Social Science & Medicine*, 55, pp. 125–39.

Marshall, M., and Phillips, D., 1999. A qualitative study of the professional relationship between family physicians and hospital specialists. *The Professional Geographer*, 51(2), pp. 274–82.

McIntyre, A., 2003. Through the eyes of women: photovoice and participatory research as tools for reimagining place. *Gender, Place and Culture: A Journal of Feminist Geography*, 10(1), pp. 47–66.

Moss, P., 1993. Research design in feminist geographic analysis. *The Great Lakes Geographer*, 1(1), pp. 31–46.

Moss, P., and Dyck, I., 2003. *Women, body, illness: space and identity in the everyday lives of women with chronic illness*. Lanham, MD: Rowman & Littlefield Publishers.

Parr, H., 2004. Medical geography: critical medical and health geography? *Progress in Human Geography*, 28(2), pp. 246–257.

Pile, S., 2010. Emotions and affect in recent human geography. *Transactions of the Institute of British Columbia*, 35(1), pp. 5–20.

Richmond, C., Ross, N. and Egeland, G., 2007. Social support and thriving health: a new approach to understanding the health of indigenous Canadians. *American Journal of Public Health*, 97(10), pp. 1825–33.

Roberts, J. M., and Sanders, T., 2005. Before, during and after: realism, reflexivity and ethnography. *The Sociological Review*, 53(2), pp. 294–313.

Robinson, K., and Elliott, S., 1999. Community development approaches to heart health promotion: a geographical perspective. *The Professional Geographer*, 51(2), pp. 283–95.

Smith, S. J., and Easterlow, D., 2005. The strange geography of health inequalities. *Transactions of the Institute of British Geographers*, 30(2), pp. 173–190.

Thrift, N., 2000. Dead or alive? In Cook, I., Crouch, D., Naylor, S., and Ryan, J. (eds.) *Cultural Turns/Geographical Turns: Perspectives on Cultural Geography*. Prentice Hall, Harlow, Essex.

Wakefield, S., and McMullan, C., 2005. Healing in places of decline: (re)imagining everyday landscapes in Hamilton, Ontario. *Health and Place*, 11(4), pp. 299–312.

Williams, A. (ed.), 2007. *Therapeutic landscapes*. Aldershot; Burlington, VT: Ashgate.

Wilson, K., 2003. Therapeutic landscapes and First Nations peoples: an exploration of culture, health and place. *Health and place*, 9(2), pp. 83–93.

Wilton, R., 1999. Qualitative health research: negotiating life with HIV/AIDS. *The Professional Geographe*r, 51(2), pp. 254–64.

Part I

Representation, ethics and power

2 Placing narrative correspondence in the geographer's toolbox

Insights from care research in New Zealand[1]

Christine Milligan

Reflections on praxis

In the last decade or so, there has been a shift in health-related research toward recognizing the importance of user/patient participation. In the UK, the Research Council's UK Public Engagement with Research Strategy (2014) has made explicit its expectation that participants will play a greater role in shaping research and having access to the knowledge generated. This shift means that more participatory approaches to research – in which the power relationships between researcher and researched are more equal – are increasingly coming to the fore. As a consequence we have seen an increase in narrative research – of which narrative correspondence forms one strand.

The shifting power balance, of course, is not without its challenges. The research agenda of the participant may not necessarily be that of the researcher. Due to the (physical) distance between researcher and researched in narrative correspondence techniques, this can present challenges. For example, in a recent pilot study on the care experiences of older male spousal carers using this approach, one respondent was more concerned about giving a detailed account, with photographs, of an invention he had developed to alleviate his wife's swollen and painful knees than he was about addressing the core themes of the study itself (Milligan and Morbey, 2013). Unlike face-to-face methods where it is possible to gently guide a respondent back to the study themes, this presents challenges for those using narrative correspondence methods.

Within the same study, I was struck by the differences in presentation of the narratives presented – from the 'scientific' account of the 'care invention' to a self-published bound book with accompanying narrative to a narrative presented in the format of a play with different scenarios relating to the 'dramatis personae' involved in the carer's life. To date, I have chosen to analyze these data thematically, however there is real potential to adopt different forms of narrative

1 This chapter is reprinted from the *New Zealand Geographer*, with the addition of the 'Reflections on Praxis' section.

and sociolinguistic analysis that engages more fully with the social, temporal (and spatial) context of the individual narrative. Though there are a number of ways of doing this, (see for example, Thomas, et al., 2009; and Phoenix, et al., 2010) essentially this involves reading and analyzing individual narratives 'as a longitudinal whole' rather than a series of themes cut horizontally across a whole data set of narrative correspondence.

Clearly how a respondent shapes their narrative speaks volumes about what is most important to them, but it is nevertheless crucial that the researcher is able to gather the data necessary to fulfill the key objectives of the study. This means ensuring that the researcher has some means of contacting the respondent for further information should it be required. The growth of technology and its widespread use has certainly simplified this, with over half of the older male carers in the study mentioned above choosing electronic format for the delivery of their narrative correspondence. A recent book on solicited diary approaches (Bartlett and Milligan, forthcoming) has also demonstrated the range of so-called vulnerable or excluded groups from a wide range of socio-economic backgrounds (including older and disabled people, carers, people with dementia, women who had experienced sexual abuse, etc.) that have successfully participated in studies using narrative and written formats. This of itself, successfully refutes somewhat frustrating, but continued, critiques – often by fellow researchers in the review process – that the reliance on the written word in narrative approaches is exclusionary and likely to result in a bias towards more affluent and educated respondents. Clearly for some groups this will not be the most effective approach, but just like any good research study, the strengths and weaknesses of a method should be weighed up with the final choice reflecting the most appropriate tools for achieving the study objectives.

Perhaps one final observation worth making on the praxis of narrative correspondence is that one very practical strength of this approach is its ease and relatively low cost of application. By this I mean that, unlike face-to-face methods that require setting up a time and venue and which involve travel *to* that venue, once initial contact has been made with those who might help identify or disseminate the call for participants, there is little by way of fieldwork to do but wait for potential participants to get in touch. This can of course be frustrating if participants do not come forward as quickly as you might like, but not more so than most other searches for participants. The added advantage is that because travel costs are far less prohibitive than in face-to-face methods, the geographical spread of narrative correspondence studies can be as large or as small as the researcher wants to make it. Similarly, the growing numbers of respondents choosing to submit their narratives electronically will in many instances cut down significantly on transcription time and costs.

Introduction

In both the title and preface to his 1995 book, Frank refers us to the 'wounded storyteller', a figure from ancient Greek mythology, whose narrative power was seen to derive from his wound.[2] In more contemporary texts, Charon (1994) wrote of the healer's need to use his or her own experiences to increase the empathic bond with those they care for, while Frank's (1995) own work gave voice to the suffering body through narrative accounts of terminal illness. Serious illness, he maintained, creates 'a loss of the 'destination and map' that had previously guided the ill person's life' (p. 1). Ill people, thus, need to 'think differently' and learn by hearing themselves tell and share their stories. In doing so, they give voice to the body – and through these stories the changed body once again becomes familiar. The power of the story, in health and illness, then, lies in its ability to not only enable the wounded to recover a voice, but to facilitate their ability to redraw the map to find new destinations. The narrative also enables others to become witnesses to the conditions that medicine cannot describe and which act to rob the wounded of their voices. Such narratives are both personal – embodied in a specific individual, and social – in that they take their narrative from the context within which they are embedded (Frank, 1995).

Narrative is not new to the geographer's toolbox. Indeed, in the last decade or so we have seen a number of geographical studies that have sought to draw on narrative approaches as a means of gaining new insights into the meanings attached to and people's everyday experiences of place (Kearns, 1997). Work in this vein has addressed such diverse issues as migration (Miles and Crush, 1993), women's desires to tan (Garvin and Wilson, 1999), service users and providers experiences of long-term care restructuring (Cloutier-Fisher and Joseph, 2000), everyday urban culture (Latham, 2003), exclusion and inclusion amongst people with learning disabilities (Hall, 2004), and health and activity patterns amongst older people (Milligan, et al., 2005). With few exceptions, however, these geographical works have used narrative as shorthand for talk. In differing ways and in differing settings, they draw on oral and aural narratives as a means of exploring the issues of interest. Geographers who have sought to engage with the written narrative as data have done so largely through documentary or archival research drawing on private (mainly historical) travel diaries (Rimmer and Davenport, 1998; Blunt, 2000; Davidson, 2002), or through the use of time-space diaries concerned with issues of travel and tourism (Thornton, et al., 1999; Gray, et al., 2001; Carr, 2002). Though a few geographers have used solicited narrative accounts as part of a wider research strategy (Latham, 2003; Milligan, et al., 2005), Meth's (2003) use of solicited diaries to examine the experience of violence amongst South African women is perhaps one of the few geographical studies in recent years that has relied predominantly on written narrative accounts as the main research tool.

In thinking about these issues, I am drawn to reflect on Thrift's (2000) call for qualitative geographers to shift away from a methodological conservatism that foregrounds techniques such as interviews, focus groups and ethnographies,

2 The wounded storyteller is based on the myth of the blind seer, Tiresius, accredited with having told Oedipus the truth of his parentage.

to be more methodologically imaginative in our research. While Latham (2003) rightly pointed out that some geographers have, indeed, sought to engage with methodological innovation, there is nevertheless a need for geographers, more generally, to think about new insights and different forms of knowledge that might be gained from a reworking of the ways in which they undertake research. This chapter represents one such attempt to do so. It discusses the extent to which narrative techniques – in particular narrative correspondence – might provide new insights into the complex relations and meanings that occur between people and place. I suggest, here, that narrative approaches open up intriguing possibilities for geographers – both conceptual and methodological. However, like all research methods, we also need to be aware of their limitations. In the remainder of this chapter, I thus reflect on these issues and draw on a small narrative correspondence study that focused on informal carers' experiences of the transition of care from the domestic to the residential care home undertaken in New Zealand to illustrate these issues.

Conceptualizing narrative accounts

Somers (1994, p. 606) argued that while the long association of storytelling with the humanities relegated the concept of narrative to 'the role of social science's 'epistemological other', over the past few decades, other disciplines have been quietly appropriating and reconceptualizing narratives as concepts of social epistemology and social ontology. That is, we constitute our social identities through narratives and narrativity; we come to be who we are, by being located, or locating ourselves, in social narratives that are rarely of our own making. 'Narratives', Somers maintained, 'are constellations of *relationships* (connected parts) embedded in *time and space* constituted by causal employment'[3] (1994, p. 616). Hence, it is by emphasizing the embeddedness of identity in overlapping networks of relations that shift over time and space that we can avoid categorical rigidities.

In discussing these issues, Somers points to four different dimensions of narratives:

> *Ontological narratives*: those social stories actors use to make sense of their lives, i.e., they are used to define who we *are* and what we *do*. These, in turn, produce new narratives and actions. The relationship between narrative and ontology is, thus, processual and mutually constitutive, but can only exist in the course of social and structural interactions over time and space.

> *Public narratives*: consisting of those stories that are attached to cultural and institutional formations that are larger than the individual (ranging from the family and workplace to the church, government or nation). Such stories are seen to have both drama and plot and derive their explanation from the

3 Employment distinguishes narrative from chronicle or annals in that it offers an accounting of why the narrative has the storyline it does, i.e., it refers to that which gives significance to independent instances, not their chronological or categorical order.

selective appropriation of events to construct a story about a specific experience/ event/ issue, etc.

Metanarratives: those stories which embed us, as contemporary actors, in a temporal and social science framework – the 'epic dramas' of our time that are built on grand abstractions.

Conceptual narratives: those concepts and explanations that we construct as social researchers.

Here, neither social action, nor institution building, are seen to be solely the products of ontological or public narratives, rather our concepts and explanations include social patterns, practices and constraints. The challenge of conceptual narrativity is to find a lexis to reconstruct and plot over time and space the ontological narratives and relationships of actors that inform their lives and the intersection of those narratives with other relevant social forces.

The ontological dimensions of narrative studies thus offer us the opportunity to extend beyond a simple methodological dimension, to engage with empirically based research into social action and agency that is spatial, temporal, relational and cultural as well as institutional and structural (Somers, 1994).

Narrative correspondence – new insights for geographical research?

The increasing attention given to narrative methods in recent years reflects a general recognition in the social sciences of the need to engage in more participatory and empowering research processes. That is, the engagement with methods which place decisions around the form of participation, the extent of the data given and the ownership of the data puts power more firmly in the hands of the researched. One such approach has been that of narrative correspondence, a technique that solicits stories from participants, written in the absence of the researcher around their experiences of particular events or processes in their lives and the meanings they attach to them. Thomas's (1998; 1999) work, for example, highlights the usefulness of this technique: firstly, for drawing out disabled women's stories about living with disability and impairment in childhood; and secondly, for illuminating some of the negative health encounters experienced by disabled women during pregnancy. Grinyer's (2002; 2004) use of this approach enabled her to uncover the often poignant accounts of parental experiences of living with young adults with cancer. Narrative correspondence thus facilitated both Thomas's and Grinyer's ability to elicit meaningful insights into the experiences of health and impairment amongst vulnerable groups. In an interesting variation of this approach, Kralik et al. (2000) engaged in more sustained correspondence as a means of exploring and sharing the ways in which women with chronic illness adapt their lives in order to tolerate their illness experience. Over a 12-month period, the main researcher corresponded regularly with participants, engaging in what amounted to an almost 'pen-pal' relationship. As a consequence of this ongoing dialogue, the emerging relationship between the researcher and the researched was seen to become one that was based on reciprocity rather than hierarchy. Such an approach can thus act to diminish the problematic of hierarchy that is often present in face-to-face methods of research with vulnerable groups. Narrative correspondence also allows participants to remain in control of the process.

In general, recruitment to a narrative correspondence study is achieved through placing a call for participants through subject-specific or publicly accessible media. In attempting to explore informal carers' experiences of the transition in the place of care in New Zealand, for example, an appeal for informal carers of older people who were in residential care, or who have recently been in residential care, was sent out through two main sources:

1 Newsletters, notice boards, e-mail and other networks known to voluntary organizations that have close contact with informal carers of older people across New Zealand; and
2 An advertising appeal placed in community newspapers that covered key geographical areas. Newspapers were selected firstly, from areas identified (from census data) as having a significant population of older people; and secondly, for the breadth of their distribution.

This approach aimed to ensure that the appeal reached as wide a range of informal carers as possible. The appeal outlined the key aims of the research and asked potential participants to get in touch with the researcher. Respondents were then sent an information sheet, consent form and some narrative prompts (see Table 2.1). Those willing to participate were asked to submit either a written or audio account of their experiences of the care transition using their own words. The decision whether or not to respond, and if so when, was thus placed solely in the hands of the respondent.

Grinyer (2004) maintains that this approach is both more anonymous and less intrusive than interview approaches, in that participants are not subject to pressure from gatekeepers to participate, nor from the researcher arriving for an interview or telephoning at some prearranged time. Respondents can decide whether or not to participate, and having done so, can contribute at a time of their own choosing and at their own pace. There is no prescription or preconceived notion about the time involved in constructing the narrative or its length – which can be as long or as short as the respondent wishes to make it. Though such narratives are inevitably written with a certain agenda in mind, they also allow room for informants to identify their own priorities. Participants have the opportunity to explore, think through, add, cut and revise their story in their own time, in their own way and with their own words, until they feel comfortable with the narrative they have expressed.

While carer respondents in the New Zealand study were given a list of narrative prompts, they were free to ignore these should they wish to do so and to write about other issues that were viewed as more meaningful and important to them. Significantly, nearly all the carer narratives highlighted the emotional distress

Table 2.1 Narrative prompts

When you write/talk about your experiences of caring for an elderly partner, relative or close friend who has moved into a residential or nursing home, you can make your account as long or as short as you like. I would like to hear not only about your experiences around the time of the move from care given at home to residential/nursing home care, but also how your experiences may have changed over time.

You are free to write about any experiences you feel are of relevance; however, I would like you to focus on the problems and issues that you and your older relative experienced during the course of their stay in the residential/nursing home. I have included a list of themes that you may find helpful to think about.

- Can you tell me a little bit about your role as a carer and who you care[d] for?
- Tell me about your experiences of dealing with the initial move from care given at home to residential/nursing home care.
- To what extent did you feel 'at home' in continuing to care for [cared for name] in the residential/nursing home?
- How were you and [cared for name] included in discussions around care and care planning for [cared for name]?
- Did you feel your own health changed as a result of your [cared for name] entering residential/nursing home care? If so in what ways?
- How do you think residential/nursing homes could adapt their practices to make carers such as yourself feel more included in the care given to those you care for?
- Please include any additional aspects of your experience of caring for [cared for name] in the nursing/residential home that you think are important.

experienced as a consequence of the care transition, ranging from worry and guilt, to feelings of fear, failure, betrayal, loss and helplessness. As Sam[4] wrote:

> The night before she went, she had dozed on our settee and had leaned over to one side. I sat beside her and told her I was sorry I couldn't look after her any more, I helped her up, took her to the toilet, got her ready for bed and helped her settle. I went and sat in the lounge and cried and cried. I felt distraught, that I had betrayed her, I had let her down, I felt guilty, I was devastated and felt desolate . . . I had an intense period of grief for a few months and although I am not a depressive person I had some of the physical symptoms.

Hence, any assumption that entry to residential care will bring mental and physical relief to the informal carer from the tasks of caring in the home does not necessarily hold true. While a few carers in the New Zealand study did, indeed, feel an immediate beneficial impact on their own health and well-being, or did so at some time shortly after the immediate period of transition, many others wrote of experiencing a range of differing health problems ranging from stress, depression and exhaustion to hypertension and high blood pressure. So while the

4 All names used in this study are pseudonyms.

transition of care may bring physical relief for informal carers who are no longer able to cope, in almost all cases, this is tempered by new concerns that affect their mental well-being.

Through the narratives, spouse carers also revealed how this transition in the place of care marked a significant change in their own life experience as they found themselves having to come to terms with lone-dwelling after many years of cohabitation. Even given the declining health of the care recipient, spouse carers noted that the shift to residential care was marked by a loss of companionship and loneliness. For some, this was described as akin to the experience of bereavement but without the finality and closure of death – and, importantly, with few mechanisms in place to support the carer through what can be a very difficult period. Spouse carers wrote of knowing that while the care recipient would never return to live in the domestic home, they felt unable to remove or pack away objects belonging to them, despite being aware that these objects would never be used again. As Anna noted, 'The hardest thing is the continual grief. I only need to open a cupboard and there may be something of Steven's and I feel devastated. I cannot dispose of all of his possessions because although he will never use them again, he is still alive.'

Narrative correspondence, then, can prove useful for capturing the meaning and weight respondents attach to different events and problems in their lives around a given issue. While most respondents broadly addressed the narrative prompts, they did so in different ways and most introduced additional issues that gave meaning to their own stories.

Of course, such narratives will vary in length, content and level of reflection. The average length of narrative correspondence in the carers' study, for example, was six pages, but this ranged in length from two to fifteen pages. Despite this difference in the length of the narratives, the end product represented a story that the *participant* felt comfortable in revealing and gave voice to that which may, otherwise, have remained silent (Frank, 1995). Jenny's narrative, for example, though relatively brief at only two pages, gave a clear account of the unhappy relationship she had experienced with her husband over many years. For her, caring represented a duty to be fulfilled rather than an extension of a shared loving and caring relationship experienced throughout their marriage. Her narrative was distinct from that of most other respondents, in that the transition of care from the home to the care home was marked, largely, by relief and a desire to distance herself from the caring process, rather than the sense of disorientation and emotional trauma expressed by most carers in the study. She wrote:

> How did I feel when my husband went into [the care home]? If I had been capable of it, I would have jumped over the moon. I was going downhill tending him and coping with [my own] heart attacks and a spleen tumor. . . . He was a heavy drinker, I lived with that for 53.5 years, now I was being told he was the one who was sick. [My husband] said I was putting on an act and the doctors believed me, if I did have a tumor it was benign. He wouldn't even get me a glass of water when I was so sick I couldn't get out of bed! . . .

> He was in for respite care, they didn't let him come home, he was in and was staying in and I was free. Not so, he rings me morning and night, always, when will I see you? Yes, the home has taken most of the responsibility for him but I still have some. I still feel I'm tied, I buy his clothes, his sweets, anything he needs.

Jenny's account revealed her anger that her own illness had been ignored, demeaned and subsumed by her husband's concern with his own ill-health. Whilst it may seem surprising that someone should reveal such deeply personal thoughts in this way, as Kralik et al. (2000) points out, there is an anonymity about writing that makes it easier to be absolutely candid than is the case with face-to-face methods. Indeed, at one point Jenny wrote of her loathing for her husband, an emotion she feared was turning to hate due to her resentment at having to take on the caring role in the home. She completed her narrative by commenting, 'Would I bring him home? Not on your Nellie! I'd go off into the wild blue yonder!' Conversely, Sam's very detailed narrative demonstrated not only the close and loving relationship he had always enjoyed with his wife, but also revealed the importance he attached to both the physical and emotional experiences of the care transition. In an excerpt from his story Sam wrote:

> Beds close to where I live were scarce but one with 20 beds was about a mile away, although at that time it was full. By October I realized I wasn't going to be able to care for Rebecca much longer but thought I would hold out until after Christmas. . . . A bed became available mid-November. That left me with the dilemma of taking it or waiting for what could be months. I felt I had no option but to take it. I felt awful, I didn't want her to go, she loved me and had cared for me all those years and I loved her and wanted to do the same for her but just couldn't.

These two very different excerpts illustrate a key strength of narrative correspondence; that is, that it places control over the data in the hands of the participants themselves, enabling them to not only consider their written responses in advance of submitting them to the researcher, but also allowing them the opportunity to reveal as little or as much as they feel willing and able to do so. Fear of disapprobation, I would suggest, means it is unlikely that Jenny would have revealed the very deep personal anger she felt about having to care for her husband in the familial home, and the sense of relief and freedom she felt following the transition to residential care, if more traditional face-to-face research methods had been employed. Similarly, it is unlikely that Sam's account would have contained the same level of personal reflection. Narrative correspondence, thus, has the advantage of encouraging informants to actively participate in both recording and reflecting on their own behaviors. Having agreed to participate, participants are able to not only edit and review their contribution until they feel happy that the narrative they send is a true reflection of their feelings and experiences, but they also have the freedom to reveal as little or as much as they feel willing or able to do so.

Written narratives of this kind also prove a cathartic experience for some research participants. Kralik et al. (2000), Meth (2003) and others have pointed to reports by participants of the therapeutic effect of writing accounts of their experiences – often acting as a 'release valve' for pent up thoughts or feelings. The therapeutic benefit seems to unfold as, through their writing, respondents develop an understanding of their own actions, emotions and experiences. In Meth's (2003) study participants refer to the narrative experience as akin to 'taking a weight off their shoulders' through the unburdening of their emotional baggage. For others, the personal stories revealed in their narratives provide a means of encouraging them to reflect on their lives and reframe them in new and more positive ways. One diary respondent in a study looking at the contributory role of activities in maintaining the health and well-being of older people wrote:

> I enjoyed doing the diaries because it made you reflect on what's been happening. You start with the feeling that – well, bugger all has happened this week, and then you search your mind and realize that, god, I did do this, I did do that or the other. Then I read it after I've completed it, and you know, I was under the impression not much had happened this week, but I've been here, there and everywhere! All sorts of things have been happening!' (Milligan, et al., 2005)

However, while research accounts refer to the therapeutic dimension of personal narratives, it is worth noting that this is not always a pleasurable experience. In the New Zealand study, some informal carers openly acknowledged the emotional journey involved in writing narrative accounts of their experiences of the transition of care for their spouse or close family member. Reflecting on his own experience, Sam wrote for example, 'You asked how I felt? That was the most difficult and emotional decision I have ever had to make in my 75 years, and now, more than two years later, I still cry as I write, remembering.' So while written narratives may have a therapeutic effect, it is also important to remain aware to the fact that reflecting on past events related to health and illness can also be a stressful and potentially a painful experience.

Finally, while one of the stumbling blocks to more traditional face-to-face methods of qualitative research has been the time and financial difficulties inherent in accessing widespread groups, this approach opens up the geographical boundaries of possible participants. Respondents may be sought from across as large or as small a geographical area as desired – a factor enhanced by widening use of the Internet and electronic mailing systems. While most participants in the carers' study came from Auckland, Dunedin and Napier, this was more a reflection of the tight timescale of the study (three months) and the three areas in which locally based managers of voluntary organizations took a particular interest in promoting the project. These managers actively raised local awareness through circulating the advert at local meetings and events, placing it in local newsletters, alerting other relevant local organizations (both voluntary and statutory) to the project and sending copies of the advert to informal carers who met the criteria for the study. The active support of these managers added weight to the study and is also likely to have encouraged participants to respond.

Despite this, as evident in Table 2.2, participants came from across both north and south islands. A less time-limited study would have enabled the researcher to focus on widening the geographical spread of participation. Table 2.2 also reveals that respondents came from a wide range of ages (from 54 to 83 years) with an average age of 72 years, thus older age is not a barrier to participation in this kind of study. While the gender balance was uneven this was not unanticipated and does, in fact, accurately reflect the greater number of female carers in the population as a whole (Milligan, 2004). Though some commentators maintain that narrative techniques are most effective when respondents are women rather than men (Keleher and Verrinder, 2003), Milligan et al. (2005) found that amongst older people, such methods can, in fact, be used to equal effect amongst both men and women.

Table 2.2 Respondent characteristics

Pseudonym	*Age*	*Cared for*	*Age*	*Illness*	*Former Residence*	*Location*
Isobel	55	Mother	81	dementia*	Non	Whakatane
Jenny	80	Husband	81	stroke	Co	Mosgiel
Maggie	54	Mother	90	dementia	Non	Pakuranga, Au.
Bren	79	Husband	87	dementia	Co	Napier
Alistair	82	Wife	77	dementia	Co	Napier
Carol	57	Mother	83	dementia	Non	Mosgiel
Susan	73	Mother	95	Rheumatoid arthritis/ heart	Non	Papakura, Au.
Elaine	63	Mother	85	Epilepsy/heart	Co	Pakuranga, Au.
Erica	62	Mother	86	stroke	Non	Howick, Au.
Anna	63	Husband	69	Parkinson's / dementia	Co	Dunedin
Norma	72	Husband	72	Parkinson's / dementia	Co	Napier
Karen	83	Husband	88	dementia	Co	Napier
Sam	75	Wife	72	dementia	Co	Napier
Libby	76	Husband	79	dementia	Co	Napier
Annie	76	Husband	93	dementia	Co	Napier
David	83	Wife	82	dementia	Co	Geraldine
Ella	82	Husband	87	dementia	Co	Dunedin
Jessica	61	Husband	72	dementia	Co	Oamaru
Rob	75	Wife	71	dementia	Co	Taradale
Brian	82	Wife	77	dementia	Co	Mosgiel

* Though it is recognized that care recipients experienced a variety of different forms of dementia, for the purposes of this study, all are referred to under the broad term of dementia.

Narrative correspondence techniques, thus, offer considerable potential for qualitative geographical research. Within the subdiscipline of health geography, for example, it may provide an additional tool for uncovering the everyday meanings and experiences of health and illness and how these experiences shape, and are shaped by, the places in which they are situated. Following transitions in the place of care from the home to residential care, for example, informal carers can feel 'dislocated' from the caregiving environment. Evidence from the carer narratives in the New Zealand study revealed how these carers sought to address this problem by seeking to forge new identities for themselves in the new care setting. Indeed, their narratives revealed that they undertook a wide range of caring tasks that fell into four main groups.

Firstly, they engaged in a range of physical care tasks – ranging from feeding, limited personal care and laundering clothing to purchasing personal items, equipment, 'treats', and so on. As Karen noted: 'Well, I feed him and give him his drink, and you know, is this the right food or would you rather have something else? And little personal things, I would have his radio in there and taking him little treats and things like that.' Secondly, they undertake a range of social care tasks, such as visiting and entertaining, taking the care recipient on outings, to family events and acting as a conduit to the outside world. As Isobel noted:

> We do *try* and get her down to our place here for special things like her first great grandchild. So we can get her together with the family and include mum and then she gets picked up and taken back to the rest home. So it's more the social, yes, those sorts of things.

Thirdly, they engage in the emotional work of caring through demonstrating affection and love to the care recipient, providing companionship and emotional support. Rob remarked, 'I make a point, with my wife, when I do see her, I always give her a cuddle and I usually smooth her hair or whatever, because I think there's some innate way we respond to that, even if we don't know.' Fourthly, they undertake an important monitoring role, checking medication and personal appearance as well as the quality of care given to the care recipient. As Erica wrote, 'One day I went to see her and she was terribly agitated. On investigation I discovered that her catheter had overfilled.' Where problems arise they also seek to ensure that these are raised with paid staff or care managers with aim of ensuring they are satisfactorily resolved. However, as Isobel's narrative revealed, informal carers can also undertake one further important role – one which has been little recognized to date – that involves helping other residents and monitoring the quality of *their* care. She noted:

> Not everybody has got family, so what I do is have one or two other people that I keep contact with when I go in. I pick up on somebody that doesn't seem to get visitors and just make a point of saying hello, and taking them for the odd walk and whatnot. So I try to go a little bit beyond my own mum.

This is particularly significant for those who have few visitors. Some carers in this study, for example, noted that while visiting their own spouse or close family member, they regularly spent time visiting and befriending other residents within the care home.

As well as the caring tasks noted above, the narratives revealed that many informal carers also have considerable knowledge and expertise in relation to the background and care needs of their spouse or close family member. Erica noted that:

> When mother was getting her urine infections, she used to get very agitated. I always knew what it was, and I would say to them, 'mother's got a urine infection' and they would said to me, 'how do you know?' and I would try to explain to them how I knew – and they'd say that because she wears a napkin, it's very difficult to take a urine sample, and it normally took about four days – and finally when it started bleeding, they would realize that she really *did* have a urine infection and she would be put on antibiotics.

Where care recipients are unable to communicate easily, the close attention informal carers pay to changes in the behavior patterns of their spouse or close family member, or their ability to recognize specific signals of distress due to their intimate knowledge of the individual, can, where acted upon, help to alleviate particular health conditions or discomfort. Facilitating informal carers' ability to take on these caring identities in the new care setting and acknowledging the important insights they can offer into the lives and care needs of the care recipients can be critical in helping paid care staff to understand how or why a care recipient may respond to situations or activities in specific ways and what may help to stimulate them. So while paid carers can offer expertise in the physical care of the individual, informal carers can offer crucial insights into the wider social, cultural and biographical background of the care recipient that are essentially an 'unknown' to paid care staff and can be important in contributing to the overall quality of care. The narratives, thus, reveal how recognition of informal carers' expertise by paid carers within the nursing home can be of vital importance in facilitating the quality of care.

How such narratives are analyzed depends on the purpose of the research. In the carers' study, all narratives were transcribed in full into a format suitable for content analysis using the qualitative software analysis program ATLAS.ti. The main themes and the complex of meanings and ideas that emerged from these data were then interpreted thematically. This was then returned to the participants in the form of a project summary, with participants being encouraged to provide feedback on their views on the emergent themes in order to ensure reliability of the data. This, of course, represents a traditional approach to the analysis of qualitative data and is one area where there is potential for geographers to respond to Thrift's (2000) call for more methodological innovation. In a recent article Wiles et al. (2005) do just that, as they call on geographers to think more creatively about how we *analyze* those texts and transcripts that form the basis of much of our qualitative data. Shifting away from thematic approaches to analysis they draw on linguistic techniques to illustrate how narrative analysis can be used to help geographers interpret and represent the complex, often fragmented, 'messiness' of people's everyday social and spatial relations. Such an approach demonstrates that it is not just the content of the stories we tell that are important, but the *kinds* of stories we tell, how they are framed, expressed and presented.

Limitations to narrative correspondence techniques

The use of narrative correspondence, then, helped to give voice to informal carers' experiences of the care transition and revealed some of the ways in which, through the creation of new caring identities in the care home setting, they sought to 'redraw the map' and find a new destination for themselves in the illness trajectory of their spouse or close family member. Of course, like all research methods, narrative correspondence has its limitations. One 'unknown' with this technique is the size of sample that will ultimately be achieved. It is, thus, impossible to quantify the sample size in advance. The New Zealand study was limited to twenty narratives due, in part, to the short timescale of the study, and in part by the timing of newsletters sent out by those voluntary organizations that had agreed to send out the appeal. Nevertheless, those carers who did respond within the timescale submitted valuable and insightful data. Given the level of interest in the subject, a longer study would almost certainly have generated a far larger sample of participants. Indeed, other studies using this approach have demonstrated that quite large samples can be achieved. Thomas (1998) for example, was able to draw on written narratives from 49 women and audio narratives from five more gathered over a period of just over one year. So with sufficient available time, this approach can be used for significantly larger qualitative studies. However, it is important to note that while, in part, the sample process can be purposive, targeting a specific group of individuals and defining particular inclusion criteria (such as age, gender, geographical location, specific experience, etc.), recruiting participants via an appeal means the final sample will ultimately be self-selected from within this group and raises the possibility of over-representation from any one particular background.

The second limitation with this method is its reliance on the written form. Clearly such a technique is unsuitable for research amongst cultural groups where there is no formal written language, or where literacy levels are poor. Similarly it may prove exclusionary for those who either find it difficult to write due to physical or cognitive disabilities or who generally feel uncomfortable with the written form. Such problems are not insurmountable, however. In the carers' study, respondents who felt more comfortable giving an oral account were offered the option of submitting either a tape-recorded audio narrative or giving a recorded audio account of their story by telephone. Both of these approaches enabled the respondents to prepare their stories in advance, whilst still maintaining their anonymity. The key difference with a telephone account is that the researcher is able to explore any additional issues of interest raised by the respondent and, unless carefully handled, may result in a narrative that is more akin to a semi-structured interview.

This raises one further issue of the comparability of the different kinds of narratives that may be submitted, from the brief, factual overview to the in-depth, reflective story. It is certainly true that those who chose to submit audio narratives tended to offer more detailed accounts of around 15 pages in length, whilst written accounts averaged around 6 pages. As Reissman (1993) pointed out, however, narratives are not exact records of what happened; rather they are representations of and reflections on past events and how they were experienced. Hence, reality will always be represented as partial, selective and imperfect. Narratives not only involve interpretation and selection in the 'telling' but also in their consumption by the researcher, as the reader, and their representation through the construction of a new story, by the researcher, out of those belonging to others. Clearly this raises questions about the epistemological status of these narratives – why some events are selected over others, what is fact and what is fantasy – and about the status of 'memories' as sources of information about life events and the nature of knowledge embedded in these narratives (Thomas, 1998). While accepting that narratives may represent only a partial and selective reality, they do offer first-hand accounts, and reality *is* being represented. Frank (1995) argues, that the 'good story' remains a self-conscious account that is truthful to life as it is lived and experienced. In health research, such accounts represent a 'pedagogy of suffering' (Frank, 1995, p. 145) that illuminates what ill people and their carers have to teach society. In this way, they can be argued to restore agency to these people. The practical problem of the narrative is to create a story in which the past is viewed in light of its connection to present and future developments. While such stories do not replace the medical narrative, they do open up the possibility of shifting between frameworks by responding to those experiencing health and illness. The insights we can gain from these first-person accounts can, thus, be of great value in gaining access to 'truths' about experiences of health and illness, as constructed by active subjects, that have a lasting meaning and effect (Thomas, 1998).

Conclusion

I set out in this chapter to consider how narrative correspondence might provide an innovative technique for gaining new and useful insights into the interrelationship between people, place and care. Drawing on sociological literatures, it seems clear that the ontological dimension of narrative approaches offers the opportunity to extend beyond the methodological to engage with empirically based social research that is spatially, temporally, culturally and institutionally aware. Thus, narrative correspondence offers considerable potential for more conceptually aware empirical research. Methodologically, it can also provide an innovative technique for uncovering the meanings people attach to the caregiving relationship and changes in the place of care. Further, as Jenny's and Sam's narratives have illustrated, the anonymity of this technique offers opportunities for respondents to reveal thoughts and feelings that they may find difficult to express in face-to-face research encounters, and which otherwise may have remained hidden. Finally, narrative techniques offer an opportunity for qualitative researchers to break down the geographical boundedness that often acts to constrain the spatial scope of their work.

I would not, however, want this discussion to be viewed as an implicit criticism of more traditional face-to-face qualitative methods, indeed, as any number of geographical studies will attest, they continue to provide valuable insights into the relationships between people and place. Rather the aim of this chapter has been to highlight one way in which qualitative geographers might usefully add to their methodological toolbox. Like all research methods, narrative correspondence has its weaknesses; nevertheless, I would suggest that, as this chapter has illustrated, it can offer geographers a useful and emancipatory method – one that meets the increasing call for more participatory and empowering research.

Acknowledgements

With the exception of the section 'Reflections on Praxis', this chapter is reprinted from 'Placing narrative correspondence in the geographer's toolbox: Insights from care research in New Zealand', in the *New Zealand Geographer*, 2005, 61, pp. 213– 224.

References

Bartlett, R., and Milligan, C., (forthcoming) *What is diary method?* ESRC research methods series, London: Bloomsbury Academic.

Blunt, A., 2000. Spatial stories under siege: British women writing from Lucknow in 1857. *Gender, Place and Culture*, 7, pp. 229–246.

Carr, N., 2002. Going with the flow: an assessment of the relationship between young people's leisure and holiday behavior. *Tourism Geographies*, 4, pp. 115–134.

Charon, R., 1994. The Narrative Road to Empathy. In Spiro, H., Curnen, M., Peschel, E., and St. James E., (eds.) *Empathy and the practice of medicine: beyond pills and the scalpel.* New Haven: Yale University Press, pp. 158–169.

Cloutier-Fisher, D., and Joseph, A., 2000. Long-term care restructuring in rural Ontario: Retrieving community service-user and provider narratives. *Social Science and Medicine*, 50, pp. 1037–1045.

Davidson, L., 2002. The 'spirit of the hills': Mountaineering in northwest Otago, New Zealand, 1882–1940. *Tourism Geographies*, 4, pp. 44–61.

Frank, A. W., 1995. *The wounded storyteller: body, illness and ethics*. Chicago and London: University of Chicago Press.

Garvin, T., and Wilson, K., 1999. The use of storytelling for understanding women's desires to tan: lessons from the field. *Professional Geographer*, 51, pp. 296–396.

Gray, D., Martin, S., Roberts, D., Farrington, J., and Shaw, J., 2001. Car dependence in rural Scotland: transport policy, devolution and the impact of the fuel duty escalator. *Journal of Rural Studies*, 17, pp. 113–125.

Grinyer, A., 2002. *Cancer in young adults: through parents' eyes*. Buckingham: Open University Press.

Grinyer, A., 2004. The narrative correspondence method: what a follow-up study can tell us about the longer term effect on participants in emotionally demanding research. *Qualitative Health Research*, 14, pp. 1326–1341.

Hall, E., 2004. Social geographies of learning disability: narratives of inclusion and exclusion. *Area*, 36, pp. 298–306.

Kearns, R., 1997. Narrative and metaphor in health geographies. *Progress in Human Geography*, 21, pp. 269–277.

Keleher, H. M., and Verrinder, G. K., 2003. Health diaries in a rural Australian study. *Journal of Qualitative Health Research*, 13, pp. 435–443.

Kralik, D., Koch, T., and Brady, B., 2000. Pen pals: correspondence as a method for data generation in qualitative research. *Journal of Advanced Nursing*, 31, pp. 909–917.

Latham, A., 2003. Research, performance, and doing human geography: some reflections on the diary-photograph, diary-interview method. *Environment and Planning A*, 35, pp. 1993–2017.

Meth, P., 2003. Entries and omissions: using solicited diaries in geographical research. *Area*, 35, pp. 195–205.

Miles, M., and Crush, J., 1993. Personal narrative as inter-active texts: collecting and interpreting migrant life histories. *The Professional Geographer*, 45, pp. 84–94.

Milligan, C., 2004. *Caring for older people in New Zealand: informal carers' experiences of the transition of care from the home to residential care*, Research Report, July, 2004. http://www.lancs.ac.uk/fss/ihr/publications/christine-milligan/nzreport.pdf.

Milligan, C., Bingley, A., and Gatrell, A., 2005. Digging deep: using diary techniques to explore health and well-being amongst older people. *Social Science and Medicine*, 61(9), pp. 1882–1892.

Milligan, C., and Morbey, H., 2013. *Older men who care: understanding their care and support needs. Final Report.* Available at: http://www.research.lancs.ac.uk/portal/en/people/christine-milligan(659b1976-a0ed-4335-b14d-14ac8e77d190)/publications.html.

Phoenix, C., Smith, B., and Sparkes, A. B., 2010. Narrative analysis in ageing studies: a typology for consideration, *Journal of Aging Studies*, 24, pp. 1–11.

Reissman, C., 1993. *Narrative analysis*. London: Sage.

Rimmer, P., Davenport, S., 1998. The geographer as itinerant: Peter Scott in flight, 1952–96. *Australian Geographical Studies*, 36, pp. 123–142.

Somers, M., 1994. The narrative constitution of identity: a relational and network approach. *Theory and Society*, 23, pp. 605–649.

Thomas, C., Reeve, J., Bingley, A., Brown, J., Payne, S., and Lynch, T., 2009. Narrative research methods in palliative care contexts: two case studies, *Journal of Pain and Symptom Management*, 37(5), pp. 788–796.

Thomas, C., 1998. Disabled women's stories about their childhood experiences. In Robinson, C., Stalker. K., (eds.) *Growing up with disability*, London and Philadelphia: Jessica Kingsley Publishers, pp. 85–96.

Thomas, C., 1999. Narrative identity and the dis-abled self. In Corker, M., French, S. (eds.) *Disability discourse*, Buckingham and Philadelphia: Open University Press.

Thornton, P., Williams, A., and Shaw, G., 1999. Re-visiting time-space diaries: an exploratory case study of tourist behavior in Cornwall, England. *Environment and Planning A*, 29, pp. 1847–1867.

Thrift, N., 2000. Dead or alive? In Cook, I., Crouch, D., Naylor, S., Ryan, J. (eds.) *Cultural turns /geographical turns: perspectives on cultural geography*. Harlow, Essex: Prentice Hall.

Wiles, J. L., Rosenberg, M., and Kearns, R. A., 2005. Narrative analysis as a strategy for understanding interview talk in geographic research. *Area*, 37(1), pp. 89–99.

3 Photo elicitation as method

A participatory approach

Tara Coleman

Reflections on praxis

Between 2008 and 2012, I undertook 'participatory photo elicitation' (PPE) interviews with older adults aged 65 to 94 to explore their experiences relating to aging-in-place on Waiheke Island, New Zealand. I had expected PPE to draw me closer to the world of each participant as it is a collaborative process. However, the closeness participants and I developed during the research was surprising and at times overwhelming. As 'the researcher' I seemed to move between an official role and a position of empathy or friendship. Ultimately, I became entangled in participants' lives, and this encouraged spontaneous, open and frank conversations during which personal matters were discussed with relative ease. But it also made me feel uneasy. I found that experiencing empathy toward participants and their circumstances could be disorienting because it caused me to feel anxious of my position in their lives and within the research (had I become a 'friend', did I want to act as an advocate for the participant, was closeness dangerous for participants?).

As I became increasingly entwined in participants' lives, I frequently wondered if I possessed a moral imperative to be responsible to some degree for supporting and assisting participants. At times I undertook activities based on a sense of moral imperative – I loaned people money, fixed broken objects, undertook housework and visited participants in hospital. In the majority of cases I carried out these activities because it was my pleasure to do so. However, on a few occasions it did not feel wholly appropriate to act as a friend. Usually this was because I did not feel sure that I would stay in touch with some participants in the long-term and did not want to place anyone at risk of being harmed by my giving a false impression of the longevity of my involvement. My perception is that some participants experienced similar confusions and were not always able to define my role in their lives clearly. This perception is based on the fact that, in communicating their requests for help, some participants seemed unsure and fearful of whether they had misinterpreted our relationship. My responses to requests varied as my emotions shifted and were not necessarily equitable. I agonized over whether this created a bias within the research. Most of all I worried about how to bring the research to a close without hurting anyone.

Some participants were distressed by the close of the research because they had enjoyed the activity of collaborating and had a clear sense of ownership over the research. To differing degrees, all participants expressed sadness that the pleasure of sharing a project would end. I asked each person if there was anything they needed to feel more comfortable about the research ending. The majority asked me to call and inform them of how the research was received. They asserted that knowing I would do so enabled them to feel considerably better. Yet it was not possible to fully protect all participants from being hurt by the end of the research. Several remained upset despite my efforts.

Feelings of worry and guilt influenced both my ability to manage my relationships with participants ethically and the manner in which I have approached the research findings. I reflected on this issue in a research logbook I maintained throughout the study:

> I have been feeling terrible about how ending the research has impacted some participants. . . . I am starting to find it difficult to know what to say when participants express a need to see me, or have strong feelings towards me. . . . Some narratives have become hard to look at. This means I have to work to manage my own emotions in order to maintain ethical relations with participants and avoid harm, and in order to see the narratives without my own emotions getting too much in the way. (September 4th, 2012)

As this excerpt shows, sharing personal information and emotions with participants may have enriched the research dialogue, but it also raised ethical dilemmas and meant that interpretation of participant narratives was a process fraught with feeling.

Despite the challenges, overall I enjoyed being immersed in participants' lives, and all participants described PPE as pleasurable and rewarding. I received genuine care and attention from participants for which I feel truly grateful. I learned much about the importance of avoiding, or at least minimizing, potential harms (to participants and researchers). I also learned that peer and supervisory support is essential in participatory and 'experience close' work. Most importantly I discovered that collaborating with participants by taking photographs together is demanding, rewarding, emotional, and potentially enriching for all involved.

Introduction

This chapter outlines 'participatory photo elicitation' (PPE) interviewing – a qualitative research strategy in which the researcher takes photographs *with* participants and *at their direction.* PPE allows participant's experiences, values and meanings related to the phenomena under study to be spontaneously shared and discussed. A detailed account of PPE is provided using illustrations from doctoral research on aging-in-place that was conducted with older adults living on Waiheke Island, New Zealand. Waiheke is situated approximately 17 kilometers east of downtown

Auckland – New Zealand's largest city which was constituted as a 'super city' in 2010. The island has a population of 8,340 permanent residents (Statistics New Zealand, 2013). The doctoral study questioned the interrelations between 'place', 'being aged' and 'well-being' that shape the experience of aging for older island residents. The key questions asked within the doctoral study were: How is 'place' and 'being aged' experienced by seniors who are ageing-in-place on Waiheke Island? How are the island and the home experienced by older island residents? And in what ways do these sites support or challenge well-being in later life?

In this chapter, I argue that PPE facilitates immersion in the world of the participant and offers opportunities to interpret the feelings, meanings and experiences that participants express during interviews, as they occur and are articulated. Further, PPE builds experiential knowledge of 'the field' and an awareness of positionality while facilitating exploration of the interconnections between place and human 'being'. Through emphasizing the *act* of photography, PPE assists wide-ranging investigation of the symbolic and material qualities of places, as well as participants' personal circumstances and the wider social contexts within which they are situated. Hence, PPE may be particularly useful for geographers and others who aim to understand the interconnections between place and human 'being'. For scholars interested in the therapeutic landscape concept, as well as those broadly focused on exploration of the dynamics between places and well-being, PPE assists understandings of the interconnections between natural, built, social and symbolic environments, allowing further insight into how these environments shape health and well-being (Gesler, 2003; Williams, 2007). PPE thus facilitates a socioecological approach by emphasizing how these interconnections inform health beliefs, practices and outcomes.

The chapter begins with a review of theoretical approaches to photo elicitation. The stages involved in the process of conducting PPE are then described. Finally, the chapter explores in detail a PPE interview undertaken with one participant, 'Polly',[1] in order to demonstrate the method and illustrate the understandings it has the potential to generate.

Theoretical approaches to photo elicitation

Photo elicitation interviewing was originally employed by Collier (1987) to study migration in response to technological and economic change. Collier found that utilizing photographs elicited more comprehensive interviews and stimulated emotional statements related to participants' day-to-day lives. Since Collier's early study, photo methods (including those that do not require the researcher's presence at the time of the photo) have been employed in a range of disciplines, including education (Rasmussen, 2004), psychology (Salmon, 2001) and geography (Smith and Barker, 2000) and with various populations. Traditionally, approaches to photo elicitation involved using photographs to instigate memory

1 Name is a pseudonym.

and discussion during a semi-structured interview (Harper, 2002). Photographs are now increasingly used in social research, reflecting growth in interest in visual culture within the social sciences (Jenkins, et al., 2008).

Photographs have long played a role in geographical studies – since the invention of photography in the 1830s, photographs have been used by geographers to illustrate 'what place is like' (Rose, 2008). More recently geographers have begun to think about photographs as 'prisms that refract what can be seen in quite particular ways' (Rose, 2008, p. 151) and are increasingly exploring how photographs may be used to understand particular contexts. Typically, geographic work does not use photographic methods in isolation but alongside interview techniques (e.g., Latham and McCormack, 2007). Geographic exploration using photographs has been extremely diverse to date, approaching topics such as outdoor adventure experiences (e.g., Loeffler, 2004), place-based experiences of living with HIV (e.g., Myers, 2010), understanding agricultural landscapes and experiences (e.g., Beilin, 2005) and children and young people's experiences of places (e.g., Rasmussen, 2004).

Photo elicitation is commonly discussed in the literature as a useful method for researchers who wish to explore participants' lived experiences (Harper, 2002). Lived experience can be an unconscious process that may be difficult to articulate, and photographs offer a way to access participants' tacit knowledge, as well as information they may be hesitant to share (e.g., information related to emotions or difficult events) (Packard, 2008). While in-depth interviewing (in all its forms) poses the challenge of establishing communication between individuals who do not necessarily share a similar background, photo elicitation methods offer opportunities to bridge gaps between the worlds of the researcher and the participant (Harper, 2002). This is because photographs (as they are viewed or produced) act as a kind of springboard for discussion and an anchor point for researcher and participant. Photo methods may therefore establish a collaborative research model (Packard, 2001). Warren (2005, p. 8) found that photo interviewing 'raises the voices of the research participants'. Similarly, using photo interviewing to explore homelessness, Wang, Cash and Powers (2000) found that participants were able to define the research agenda and discuss issues that the researchers would not have thought of, thereby highlighting participants' own perspectives. This is not to suggest that photo elicitation inevitably reduces the power balance between researcher and participant since it can be employed without regard for power dynamics (Harper, 2002). Yet photo methods possess the potential to produce equitable power relationships more so than traditional positivist approaches in which the researcher tends to be positioned as a detached 'expert' (Cancian, 1992). For this reason, researchers aiming to alter the research process in order to empower those under study and enhance their participation have looked to photo methods, especially those that establish the researcher and the participant unequivocally as collaborators.

Packard (2001, p. 64) claims that researchers who turn over the camera to the participants themselves make the 'most dedicated efforts to bring participants into the research process as co-collaborators'. However, like other interview strategies, photo elicitation approaches that require the participant to take photographs that will be discussed at a later point (e.g., 'photo voice') have disadvantages when the research is concerned with investigating the role of place in experience. If the researcher is not present when the photograph is taken, sitting down to discuss an image latterly maintains a distance between the participant and the environment under study. Hence, aspects of lived experience may be invisible or unfathomable since what is recorded in an image, as well as what is discussed about an image, remains *at a distance from where experiences have occurred.* Meanwhile, interviews in which the researcher and the participant take photographs together may allow participant's meanings and experiences to be explored and interpreted *as they interact within their environments*. The process of researcher and participant working together to produce photographs allows the researcher to 'be there' to observe participants spatial practices, gestures, expressions and the wider social setting. In sum, a participatory approach to photo elicitation may enhance an individual's agency and participation in the research while allowing sensitivity toward their environmental experiences.

Participatory photo elicitation

An interpretive approach

My decision to develop a participatory approach to photo elicitation was driven by two research goals. First, I sought to uncover the interrelations between 'place', 'being aged' and 'well-being' that shape the *experience* of aging-in-place for older adults living on Waiheke Island. Therefore, it was appropriate to adopt a phenomenological approach. Phenomenology highlights human 'being' as a matter of interdependence with other beings and everyday settings, illustrating that place is irrevocably entwined with human experience (Casey, 1993). There are many approaches to phenomenology; in developing a participatory approach to photo elicitation able to highlight experience and human 'being' as a matter of 'being-in-place', I have drawn on an interpretive approach. Interpretive phenomenology is a process and method for uncovering 'deep' information and perceptions in order to reveal human experience and human relations (Lopez and Willis, 2004). Through interpretive approaches, meanings that may be embedded in the practices and places of everyday life may be explored (Steeves, 2000). This includes meanings that may not be readily apparent to participants themselves, yet are (re)produced through participants' actions (spoken and unspoken). Hence, an interpretive strategy reveals participants' lived experiences through the illumination of participants' meanings and perceptions (Steeves, 2000). Taking an interpretive approach to PPE has meant emphasizing understandings related to what people experience, rather than focusing solely on what they consciously know.

Enhancing older adults' participation and agency

The second research goal that informed my decision to develop a participatory approach to photo elicitation was that of maximizing older adults' participation and agency within the research process. In order to meaningfully understand older adults' experiences, space must be made for their own accounts of that experience. Enhancing older adults' participation in the research means it is more likely that the research will reflect older adults' own perspectives and experiences. This is important in the current context of widespread ageism in society. While the academic literature has recognized that being aged is not inevitably a matter of being in decline and is diversely experienced (e.g., Wiles and Allen, 2010), stereotypical discourses assert being aged as being frail, unwell, dependent, incompetent, or senile (Hurd, 1999). Through researching experiences of being aged I have sought to move beyond taken-for-granted notions related to aging by enhancing older adults' participation within the research process, thereby empowering their own accounts of being aged. I have approached maximizing older adults' participation in the research as involving participants in the act of photography itself to allow their choices, feelings, values, meanings and experiences to be given space as they are spontaneously shared and discussed during the research encounter. Further, participants' agency within the research process is increased by allowing individuals to work with and direct the researcher.

Stages of participatory photo elicitation

Gathering the data

Primary recruitment of participants for the doctoral study took place at two key Waiheke Island agencies involved in senior care and support services – the Waiheke Health Trust and the Piratahi Hau Ora Trust. Other recruitment techniques took place at the local community centers on the island by making brief presentations of the study to older adult groups. The majority of participants (62%) were recruited through Waiheke Health Trust, while 16% were engaged through Piratahi Hau Ora and 22% were recruited through community centers. The criteria for inclusion in the study were: 65 years of age or over, living at home in the Waiheke community, able to communicate and understand basic English, and willing to take part in the study. Twenty-eight participants (20 women, 8 men) aged 65–94 took part in the study. All participants completed the first stage of the research – in-depth interviews – while 11 (of the 28) participants subsequently engaged in PPE interviews. These 11 participants (8 women and 3 men) were those who were willing and able to offer continued participation in the research. Though a relatively small group, participants were very diverse in terms of their length of residence on the island (ranging from 1 to 55 years). All PPE participants initially expressed anxiety about using a digital camera but were relieved to hear I would be assisting in the operation of camera equipment.

In-depth interviews were first conducted and acted as 'opening conversations' with participants. The PPE interviews that followed this first stage of the research drew on themes and issues arising in the initial interviews. Throughout the research process I maintained a researcher's logbook to facilitate reflection on my developing understandings and my influence upon the research. These stages of the research process are explained in detail in the following subsections.

Initial interviews

The initial in-depth interviews employed a standardized interview template that asked open questions such as 'What is it like to live here?', 'What do you like to do here?' and 'How do you feel about your home?' These questions elicited a wide range of responses related to each participants' housing circumstances, perceptions of home and place, and activities beyond home. Through use of a standardized template, it was possible to identify themes and issues illustrative of older adults' experiences of aging-in-place on Waiheke Island for further exploration during the PPE interviews that followed. All participants showed me around their homes during initial interviews without prompting. Receiving these 'tours' allowed me to get to know the layout of participants' homes and prompted discussion of a range of feelings related to aging-in-place. Thus, initial interviews acted as 'opening conversations' (moving beyond the interview template) and softened the boundaries between my world and the worlds of participants by enabling participants and me to get to know each other and begin to share the research process. These initial interviews were digitally recorded to facilitate verbatim transcription.

Participatory photo elicitation interviews

During PPE interviews I took photographs *at the direction of participants*. To begin this process I discussed with each participant themes and issues arising in the earlier in-depth interviews (specific to each individual participant and the collective data). As these discussions progressed, I asked participants if there was anything they would like me to photograph – for example, objects, spaces, possessions or views (O/S/P/V) – to illustrate their comments. Every time I asked this question, each participant responded by selecting one or more O/S/P/V to photograph. As participants made their selections, I watched for bodily movements and gesticulations, noticed where and how participants 'set the scene' for the photograph to be taken and listened to their spoken words as they did so. This included paying attention to how participants touched, gazed upon, engaged in moments of silence with, spoke to, talked about and positioned the O/S/P/V that they had selected to photograph. Interpretation therefore unfolded during the research encounter, as well as taking place afterwards. I explored participants' choices of what to photograph and why by asking questions as participants made their selections and by discussing with participants how to arrange and produce the photographs. These questions included, but were not limited to, a series of prompts, as detailed in Table 3.1.

During PPE interviews I gave participants as little direction as possible and focused on asking questions in relation to their choices of what to photograph and how. Only one of the 11 participants was initially reticent to begin and insisted

Table 3.1 PPE interview prompts

Prompts for beginning photos
Is there anything you would like to photograph to illustrate your comments?
How would you like me to photograph this (O/S/P/V)*?
Where/how would you like this photograph arranged?
Prompts for discussion while taking photographs
What is the significance of this (O/S/P/V) in your day-to-day life?
What does this (O/S/P/V) mean to you?
How does this (O/S/P/V) make you feel?
Does this (O/S/P/V) remind you of anything or anyone?
Does this (O/S/P/V) assist you in any way?
What do you think you might do with this (O/S/P/V) in the future?
Is there any particular reason why you have chosen this location for this (O/S/P/V) to be photographed?
Is there any reason why you have arranged these (O/S/P/V) in this way?

* O/S/P/V: object/space/possessions/view

on instructions. I was able to engage this participant in taking photographs by pointing to an object we had discussed earlier in the initial interview as something important to their experience of aging-in-place and encouraging further comment on the significance of this object. Once the first photograph was taken, typically several others were constructed in a steady and comfortable stream. (Photographs were not shown to participants during this process but later were shared in a participant journal designed to check and discuss my developing interpretations.) I made jottings in my notebook whenever I could between handling equipment and conversing with the participant. However, this was sometimes challenging and frequently impossible as the participant and I were engaged in constructing a photograph while conversing. Therefore, I digitally recorded these conversations with permission and later transcribed them verbatim. While there was much to do during PPE interviews – taking photographs, observing gestures and the environment, gauging emotions, responding to the conversation – rather than create distance between myself and participants, managing the recorder and camera equipment helped to facilitate a shared experience within the research encounter. This occurred since participants would usually help me to carry around equipment. In this way participants became unexpected 'field assistants', which seemed to make it easier and more comfortable for them to direct the photographic process and for the conversation to flow. This process was demanding insofar as I was required to pay conscious attention to building rapport and facilitating empowerment for the participant. Most participants became commanding of the process and appeared empowered by the end of PPE interviews. Immediately following each encounter, I expanded on any jottings I had managed to make, noting participant's non-verbal signals (e.g., gestures, pauses, expressions) and recorded my own impressions, feelings and the overall experience of witnessing participants' acts of choosing, arranging and discussing what was photographed within their homes.

Researcher's logbook

Undertaking qualitative research is always a relational, emotional and reflective practice, which requires critical consideration of the researchers' experiences of 'being-in-the-research' (Davies and Dwyer, 2007). In the context of PPE, during which I was clearly positioned as researcher and co-creator in the act of taking photographs, it was important that I reflect on my experience of 'being in the research' explicitly throughout all stages of the research. Thus, I maintained a researcher's logbook in order to acknowledge the research as dependent upon and complicated by my own emotions, understandings, actions and experiences. In this journal I reflected on issues of positionality by examining my personal assumptions, goals, perceptions, and shifting subjectivities. I recorded my feelings, thoughts and responses at regular intervals and explored how my own perspective might have been involved in the play of each research encounter.

Interpretation of the data

A research strategy that involves in-depth interviews, PPE interviews and a researcher's logbook generates a substantial data set. To store and initially organize the data I utilized the NVivo data analysis package (McNiff, 2000). After managing the data, I openly read participants' narratives, jottings related to each research encounter and my researcher's logbook, and continued to reflect within my logbook, on many repeated occasions, in order to understand participants' situated perspectives and personal meaning-making (Smith and Osborn, 2007). I then asked the following questions of the data: 'What is happening in this account?', 'What is the meaning?', 'What does this say about place and being aged?', 'What is missing/what has not been said?', 'What are my assumptions?' and 'How does this relate to the data as a whole?' (Wright-St Clair, 2008). These questions allowed me to critically weigh participants' narratives and were themselves a way of moving forward – beyond words themselves to illuminate taken-for-granted and everyday phenomena. In this sense I approached participants' narratives as phenomenological stories and interpretation as a process of asking questions in order to be drawn deeper into hermeneutic engagement. As Wright-St Clair notes (2008), this approach facilitates focused attentiveness and assists transparency in articulating interpretive choices. The process of asking questions of the data was therefore a way of practicing and presenting a dialectical conversation that moved beyond participants' words to uncover hidden meanings within (Diekelmann, 2005).

After summarizing my initial understandings I returned once more to the data to undertake systematic coding. Transcripts and my research logbook were read again several times and detailed notes identifying key themes and experiential postures were made (Smith and Osborn, 2007). During this 'close reading', I paid particular attention to things that matter to each participant (e.g., objects, events, relationships, concepts, processes and so on), the meanings relating to these things of import and the participant's stance (Larkin, et al., 2006). I then reflected upon how it might feel and what it might mean for each participant to possess their particular experiential claims, concerns and understandings (Smith, 2004). Following this I sought emergent patterns, commonalities and differences within and across the coded data (Eatough and Smith, 2008). This included making detailed notes in the margins of transcripts and more 'free writing' in my research logbook. I also employed 'progressive focusing' (Wolcott, 1994, p. 18), which involved zeroing in on a particular detail or viewpoint expressed by a participant and then stepping back to consider other participants' perspectives and the broader context. My developing understandings were checked by discussing my interpretations with participants (which involved visiting participants to share my initial understandings) and others (e.g., supervisors). This reflection and communication about my interpretations assisted me in developing an interpretive 'story' for each participant, as well as across the data as a whole.

An example: Polly's story

Within the bounds of this chapter, it is necessary to focus on one participant's story in order to adequately illustrate PPE and explore its potential offerings. Hence, I have selected one participant's story for discussion – Polly's. During my initial interview with Polly (aged 83), she explained that she had lived in her present home on Waiheke for 16 years. She also explained that she had numerous health concerns that meant she was frequently house-bound and relied on a walker at home, as well as a wheelchair on the rare occasions that she left the house. As previously stated, open-ended questions used in initial interviews frequently prompted participants to spontaneously provide me with a tour of their homes. When Polly showed me around her home during our first conversation, she indicated that her bedroom window and living room were of particular importance to her experience of being at home as she grew older. She explained that 'spending time just looking' at views from the window and 'enjoying just sitting' in the living room was part of her daily routine and facilitated 'respite from (my) health issues', as well as 'enjoyment in being alone'. A key theme that arose from this first conversation with Polly related to her concept of home as 'a place to feel solitude but not loneliness, where it is possible to create a spiritual and healing place, (and take) a journey of renewed health'. When I returned to Polly to conduct PPE interviews it was possible to discuss this theme in more depth. To begin the PPE interview with Polly I first returned to discussion of the significance of views from her windows, as follows:

Tara: Last time I was here we were discussing the windows and views.
Polly: Yes, well, I feel quite possessive of these views, they're so important to me, so it's hard to explain.
Tara: Could you tell me about the significance of the windows and views in your life?

Polly: Maybe the best thing is to show you.
Tara: That would be great, thank you, and would you like us to take a photo of a particular window or view?
Polly: Yes, we can give it (taking photographs) a go.

As these notes show, Polly felt explaining the significance of her windows/views would be difficult but was open to communicating their meanings by experiencing them with me (i.e., showing me the window/view). Polly's openness made it relatively easy for me to suggest we take a photograph together. She immediately led me to a large window situated in her bedroom (see Figure 3.1, opposite) and we commenced discussing and photographing this window and the view below:

Polly: So, this is a window that means a lot to me. But what it means might be more than I can really explain. (Falls quiet).
Tara: Is there any particular way you imagine we could photograph this window?
Polly: Well, I think a photograph of this window should have me in it (laughs). Because I should be looking at that view out there (points to view through window).
Tara: Can you tell me why you would like to be looking at the view out there for the photograph?
Polly: Well, Dear, because that's the whole experience. It's how I take my 'therapy' here.
Tara: Can you tell me more about what you mean by that, by that word 'therapy'?
Polly: Yes, but let me show you then you'll really get it. I will lie here on the bed like this (lies down). See, it's lying here like this and looking out through (the window). That's what is special.
Tara: Can you tell me how you feel as you lie there and look out through the window?
Polly: Yes, well this is just very therapeutic. I can connect with the island out there (points to view of coastline) and the ever-changing light through this window coming inside. It's like connecting with God, I feel a certain sense of spirituality here. (Pauses). Have you taken the photograph? I will look out like this and you take the photograph.

The photograph that Polly and I produced during this discussion is illustrated in Figure 3.1. When I asked Polly what we could title this photograph she replied, 'Call it just 'Polly' because it really depicts me, the essence of my life here, Dear.'

After we had finished taking this photograph, I felt very connected to Polly. Indeed, we were in the intimate space of her bedroom while she lay in bed and discussed her feelings and experiences in relation to her practices at home, her health and her spirituality. Taking the photograph together allowed me to witness Polly in her bedroom interacting with the view through her window as she described

Figure 3.1 Polly

the significance of this act to her everyday life. Being there together to take the photograph seemed to break down any barriers between us. After the photograph (Figure 3.1) was taken, Polly talked spontaneously and with ease (without questions from me) about the therapeutic experience she had when sitting in bed and looking out through the window:

> I do spend a lot of time here in bed. . . . I listen to the radio propped up and take my medicine, just like this (demonstrates). It isn't that bad, Dear, as I've said I am blessed with the window to just gaze out through and the magical view and the light coming through. That's why this is a healing place. If all I have to worry about is being in bed when I have that lovely window to look through, then what's the problem really.

The comforting nature of being at the window, experiencing the light and feeling blessed by the views of the island beyond is evident in Polly's narrative. Her words illustrate the therapeutic qualities of her home and its role in managing times of ill health, which become less problematic when compared to the opportunity for Polly to enjoy the views afforded by her window. Polly's explanation of her choice to title this photograph simply 'Polly' (to reflect who she is at home) highlights the importance of the window, the view beyond and the act of communing with these material elements which symbolize a way for Polly to cope day-to-day. Photographing Polly at her window allowed me to be with her as she enacted and described this practice of being in bed, enjoying the view and coping with her health issues. Consequently, I felt immersed in Polly's therapeutic experience of her home since I too was experiencing the window, the view and the light, and she seemed to encourage this immersion:

> Sit here, Dear, can you feel the light flowing through the room. All this light, it's like my spirituality and faith when I'm unwell or in a bit of discomfort. It comes flooding through and I say to myself 'it's God here at my window' and I feel I can cope.

The agency to direct me that Polly felt during the interview is apparent in this comment. By urging me to share the view and light with her, Polly allowed me to share a personal experience and communicated to me clearly that having the opportunity to connect with the spirituality of light and God was important for her in times when she was in bed feeling unwell. Thus, I gained a sense of how Polly interacted with her home and how this interaction informed both her experience of being aged and her well-being. Additionally, the act of photography allowed me to view in person, rather than second-hand (e.g., through a photograph Polly might have taken herself to show and discuss with me) the materiality of Polly's home. I could experience the size, shape and position of Polly's window, its relationship to the bed where Polly spent large amounts of time and the view it offered her. I was able to directly experience the symbolic significance of Polly's home as 'healing place' as she enacted and invited me to take part in her practice of being in bed witnessing the view from the window. By positioning me as the researcher *and* co-creator in the process of documenting Polly's daily life at home, PPE interviewing thus allowed me to build experiential knowledge of 'the field'. I witnessed the sense of calm and peace Polly expressed through words taking place through emplaced bodily expressions and movements (e.g., her serene face and stillness as she gazed from the bed through the window). I also gained a sense of Polly's personal circumstances and wider social context as she continued to direct me in taking photographs, as follows:

Polly: Would you like to have a look at something else I find very helpful in my life, Dear?

Tara: Yes, thank you, I would.

Polly: Well, if I'm truthful, there are so many things in this house that have a meaning. In the main (living) area there are a lot of special things . . . everything has a little meaning and tells a little story about my life.

Tara: Would you like us to photograph the living area and talk more about that?

Polly: Yes, because it's all about things I've collected in there, things everywhere. Come and see.

(We move from the bedroom to the living room.)

Tara: How would you like this photograph taken? Is there a particular place you would like me to stand to take the photograph or an angle on this room?

Polly: Try to get it all in Dear because it's the lot of it that's important. It's not one thing, it's all of it together . . . Get (photograph) this old chair . . . this old chair was 20 dollars, the table there about 30 dollars, I have always had very little money and had to find gems among junk. So everything has a little story . . . a journey I went on to find things and how I made things work. And you feel filled up with the memories and the meaning of things, and content. So, yes, I feel very blessed here. I think that when you've been independent you're inclined to be thinking that you are alone. . . . I've come to the realization that there's a difference between isolation and solitude. And so I think that once you realize that you're part of humankind. Yes, that's right. Can you get it all in (the photograph) from there?

Tara: Yes, I think so.

Polly: Good, get it all in (laughs), it all means a lot. You see I'm not that close to my family, Dear. The family are a bit dysfunctional (laughs). I never got on with my parents, no siblings or anything. And I never found love or got married. I get a few visitors here, not many, one or two. If people want to take me out they have to deal with (my) wheelchair. So, it's mostly just me. But I feel really good here, really

> in the swing of it (life) here, I enjoy all these things and remember all the adventures I've had finding all this junk, putting my life together.

The photograph of Polly's living room that resulted from this conversation is illustrated in Figure 3.2 and was given the descriptive title 'Polly's living room' by Polly herself. As the interview notes reveal, during the process of guiding me to take a photograph of her living area Polly was able to emphasize the importance of all her possessions kept there in instigating a feeling of being surrounded in meaning and connection to the world. Through directing me to include the old chair and the table in the photograph, Polly spontaneously shared with me something of her personal circumstances and wider social context. She quite comfortably explained that she had faced financial challenges, disconnection from family relationships, had not experienced having a significant other and spent much time alone at home with few visitors. Polly was also able to clarify that by finding enjoyment in the things kept in her living area, which represent how she has coped in the context of these challenges, it was possible to see herself as part of humankind and content. This theme of 'being alone but not lonely' was articulated by all 11 PPE participants and is a prominent theme within ageing and health literature (e.g., Iecovich, et al., 2011).

As Polly moved around to direct the photograph and explain the significance of her possessions, I was able to interpret how particular objects in Polly's living area signaled particular biographies that connected her to meaningful memories and past events (e.g., finding gems among junk and coping with financial and social limitations). I immediately understood that the story of where an object was discovered, why an object was purchased, or the social and material conditions of life within which an object was found and bonded with, provided context and meaning to Polly's life at home, enabling her to feel surrounded in the meaning of things.

Moving through to the living room with Polly to take the photograph whilst operating camera and audio equipment, as well as talking with and observing Polly, was somewhat challenging. In my logbook after the interview I recalled 'feeling overwhelmed now and then' by the process. Polly agreed during the interview that 'all this is a busy business.' Yet, my interaction with Polly did not feel hurried or disconnected. Like other participants, Polly asked if she could help with equipment and seemed to feel an increased sense of agency to direct me, assist me and share her feelings with me by becoming a 'field assistant'. At one point she exclaimed: 'Oh I am enjoying this!' Thus, while disparity remained a feature of the research encounter by virtue of my official status as 'the researcher' – a status that positioned me as 'expert' and therefore afforded me intellectual power and authority within the encounter – Polly appeared empowered by the process at specific times.

When recording my overall impression of the PPE interview with Polly in my logbook after leaving her home, I described the method as 'a process of getting to know Polly's world and being entangled in the research process'. This feeling of immersion and entanglement enabled me to recognize 'the field' as a relational

Figure 3.2 Polly's living room

space (Parr, 2001). In other words, PPE interviews highlighted the mutually interactive dialogue and process of the research. In this sense the research felt personal. For example, Polly commented that through taking photographs together we had shared 'something special, very personal', and were now friends while I felt a closeness to and empathy for Polly that was akin to a feeling of friendship. For many months after our PPE interview Polly called me because she wanted 'to chat to a friend' and this was typical of all my interactions with PPE participants.

In order to navigate the feelings (mine and those of participants) that being immersed in the world of participants raised during the research, supervisory and informal support networks (comprised of friends, family members, and colleagues, for example), as well as free writing in my logbook, were essential. These support systems afforded consistent opportunities for critical reflection and problem solving. Given that the emotional aspects of research are often undervalued within the university culture, informal support networks may be of particular value. While talking to other students, friends and family members supported me in negotiating the shifting boundaries of the field, what helped most was discussion with other researchers within the academy who have had similar experiences of 'being in the research'. I also consistently reported and checked my developing understandings with participants themselves and scrutinized the research process by returning again and again to the gathered text and to my research logbook to check and recheck my interpretations and decision making. All 11 participants gave generous time and effort in considering and responding to my developing understandings. The majority affirmed that my interpretations 'made sense' and were 'vibrant, doing justice to getting older as changing but still responding to life' as one participant, John (aged 89), put it. Two participants wrote long letters to ask me to use the research to emphasize that aging is about 'being active, we're not "has-beens"', as Linda (aged 65) commented.

PPE interviews lasted several hours beyond the one hour that had been suggested as a time frame in the participant information sheet and consent forms. While participants typically did not want the research encounter to end, I used my discretion and ended the encounter if the participant appeared at all fatigued. Three participants did appear tired during PPE interviews, which prompted me to undertake several shorter visits with these individuals. I did not anticipate and was not prepared for the length of PPE interviews and was frequently exhausted afterwards myself. Future practitioners of PPE should adequately consider its demands and explain these to prospective participants. Future practitioners should also present ethical committees with realistic time frames. In sum, it is important to build a strong foundation for PPE through consideration of its demands and associated ethical issues and by creating information and consent forms that are accurate.

Conclusions

PPE assists wide-ranging investigation of the symbolic and material qualities of places, as well as participants' personal circumstances and the wider social contexts within which they are situated. Through emphasizing connections between physical, symbolic and social landscapes, PPE assists a socioecological approach to health and well-being. Hence, PPE may be particularly useful for geographers and others who aim to understand the interconnections between places, human 'being', health and well-being with respect to experiences of aging, but also in relation to a range of other phenomena, contexts and diverse social groups.

PPE maximizes opportunities for agency and participation in the research process for participants by allowing them to work with and direct the researcher. It also allows researchers to acknowledge the field as a relational space where their 'footprints', feelings and experiences matter. Moreover, PPE generates data that is salient to those involved in the research, thereby increasing the validity and quality of research outcomes. Through positioning participants as 'co-creators', PPE is inclusive and may benefit groups experiencing inequalities, as well as those historically cast as 'subjects'. Accordingly, PPE offers much to researchers, policy makers, health promoters and others who aim to improve and empower the health, well-being and social position of specific groups (including but not limited to older people). Nonetheless, PPE involves building relationships and emotions that can disorient the researcher and participants and must be managed carefully (e.g., through use of a reflective research log book, peer and supervisory support and appropriate research and ethical procedures) to avoid harms.

References

Beilin, R., 2005. Photo-elicitation and the agricultural landscape: 'seeing' and 'telling' about farming, community and place. *Visual Studies*, 20(1), pp. 56–68.

Cancian, F., 1992. Feminist science: methodologies that challenge inequality. *Gender and Society*, 6(4), pp. 623–642.

Casey, E., 1993. *Getting back into place: toward a renewed understanding of the place-world*. Bloomington: Indiana University Press.

Cobb, A., and Forbes, S., 2002. Qualitative research: what does it have to offer to the gerontologist? *The Journal of Gerontology, Series A: Biological Sciences and Medical Sciences*, 57(4), pp. 197–202.

Collier, J., 1987. Visual anthropology's contributions to the field of anthropology. *Visual Anthropology*, 1, pp. 37–46.

Davies, G., and Dwyer, C., 2007. Qualitative methods: are you enchanted or are you alienated? *Progress in Human Geography*, 3(2), pp. 257–266.

Diekelmann, J. 2005. The retrieval of method: The method of retrieval. In Ironside, P. M. (ed.), 2005. *Beyond method: philosophical conversations in health care research and scholarship*. Madison, WI: University of Wisconsin Press, pp. 3–57.

Eatough, V., and Smith, J., 2008. Interpretive phenomenological analysis. In Willig, C., and Stainton-Rogers, W. (eds.) 1980. *The Sage handbook of qualitative research in psychology*. London: Sage, pp. 179–194.

Epstein, I., Stevens, B., McKeever, P., and Baruchel, S., 2006. Photo Elicitation Interview (PEI): Using photos to elicit children's perspectives. *International Journal of Qualitative Methods*, 5(3), pp. 1–9.

Gesler, W. M., 2003. *Healing Places*. Lanham, MD: Rowman and Littlefield.

Gilgun, J., 2005. 'Grab' and good science: writing up the results of qualitative research. *Qualitative Health Research*, 15(2), pp. 256–262.

Glaser, B., and Strauss, A., 1967. *The discovery of grounded theory*. Chicago: Aldine.
Haraway, D., 1988. Situated knowledge. *Feminist Studies*, 14, pp. 575–599.
Harper, D., 2002. Talking about pictures: a case for photo elicitation. *Visual Studies*, 17(1), pp. 13–25.
Hurd, L., 1999. 'We're not old!': older women's negotiation of aging and oldness. *Journal of Aging Studies*, 13(4), pp. 419–439.
Iecovich, E., Jacobs, J., and Stessman, J., 2011. Loneliness, social networks and mortality: 18 years of follow-up. *Aging and Human Development*, 42(3), pp. 243–263.
Jenkins, K., Woodward, R., and Winter, T., 2008. The emergent production of analysis in photo elicitation: pictures of military identity. *Forum: Qualitative Social Research*, 9(3), pp. 95–106.
Kusenbach, K., 2003. Street phenomenology: the go-along as ethnographic research tool. *Ethnography*, 4(3), pp. 455–485.
Larkin, M., Watts, S., and Clifton, E., 2006. Giving voice and making sense in interpretative phenomenological analysis. *Qualitative Research in Psychology*, 3, pp. 102–120.
Latham, A., and McCormack, D. P., 2007. Digital photography and web-based assignments in an urban field-course: snapshots from Berlin. *Journal of Geography in Higher Education*, 31(2), pp. 241–256.
Lee, H., 2008. The shadow side of fieldwork: exploring the blurred borders between ethnography and life. *Qualitative Research Journal*, 8(2), pp. 150.
Loeffler, T. A., 2004. A photo elicitation study of the meanings of outdoor adventure experiences. *Journal of Leisure Research*, 36(4), pp. 536–556.
Lopez, K. and Willis, D., 2004. Descriptive versus interpretive phenomenology: their contributions to nursing knowledge. *Qualitative Health Research*, 14(5), pp. 726–735.
McNiff, K. (2000, October 7), What is qualitative research? QSR International, The NVivo Blog, October 7, 2015. Available at: http://www.qsrinternational.com/blog/What-is-Qualitative-Research.
Myers, J., 2010. Moving tales: postcards of everyday life living with HIV in Auckland, New Zealand. *GeoJournal*, 44(4), pp. 453–457.
Packard, J., 2008. 'I'm gonna show you what it's really like out here': the power and limitation of participatory visual methods. *Visual Studies*, 23(1), pp. 63–77.
Parr, H., 2001. Feeling, reading, and making bodies in space. *The Geographical Review*, 91, pp. 158–67.
Rasmussen, K., 2004. Places for children – children's places. *Childhood*, 11(2), pp. 155–173.
Rose, G., 2008. Using photographs as illustrations in human geography. *Journal of Higher Education*, 32, pp. 151–160.
Salmon, K., 2001. Remembering and reporting by children: the influence of cues and props. *Clinical Psychology Review*, 21, pp. 267–300.
Smith, J., 2004. Reflecting on the development of interpretative phenomenological analysis and its contribution to qualitative research in psychology. *Qualitative Research in Psychology*, 1, pp. 39–54.
Smith, F., and Barker, J., 2000. Contested spaces: children's experiences of out of school care in England and Wales. *Childhood*, 7(3), pp. 315–333.
Smith, J., and Osborn, M., 2007. Pain as an assault on the self: an interpretative phenomenological analysis. *Psychology and Health*, 22, pp. 517–534.
Statistics New Zealand, 2013. *Census Usually Resident Population Counts*. Available at: http://www.stats.govt.nz/browse_for_stats/population/census_counts/2013CensusUsuallyResidentPopulationCounts_HOTP2013Census.aspx. (Accessed January 15, 2015).

Steeves, R., 2000. How to start fieldwork. In Zichi Cohen, M., Kahn, D., and Steeves, R., eds. 2000. *Hermeneutic phenomenological research: a practical guide for nurse researchers*. Thousand Oaks, CA: Sage. pp. 25–35.

Wang, C., Cash, J. and Powers, L., 2000. Who knows the streets as well as the homeless? promoting personal and community action through photovoice. *Health Promotion Practice*, 1(1), pp. 81–9.

Warren, S., 2005. Photography and voice in critical qualitative management research. *Accounting, Auditing and Accountability Journal*, 18, pp. 861–82.

Wiles, J., and Allen, R., 2010. Embodied ageing in place: what does it mean to grow old. In Chouinard, V., Hall, E., and Wilton, R. (eds.) *Towards enabling geographies: 'disabled' bodies and minds in society and space*. Surrey, England: Ashgate Publishing. pp. 217–236.

Williams, A., 2007. Introduction: the continuing maturation of the therapeutic landscape concept. In Williams, A. (ed.) *Therapeutic landscapes*. Aldershot, England: Ashgate Publishing. pp. 1–12.

Wolcott, H. F., 1994. *Transforming qualitative data: description, analysis, and interpretation*. London: Sage Publications.

Wright-St Clair, V. A., 2008. *'Being aged' in the everyday: uncovering the meaning through elders' stories*. PhD thesis, University of Auckland.

4 Ethics and activism in environment and health research

Sarah A. Mason, Chad Walker, Jamie Baxter and Isaac Luginaah

Reflections on praxis

This chapter features fieldwork conducted by Mason and Walker that was ultimately challenged by community activists, so in the interest of brevity, their reflections on praxis are the focus here.

Sarah Mason

My dissertation research[1] examines perceptions of a controversial waste processing facility that was proposed for the rural community in which I grew up. Here I discuss how my role as an insider impacted me emotionally, professionally, and personally.

It was disheartening that in an effort to censor my research, local activists directly contacted our Research Ethics Board (REB) rather than merely asking me what my intentions were with the research. Some of these people knew me as a child and watched me grow up volunteering in our community. It was very stressful and upsetting to feel that they either honestly believed that I was corrupt, working for government and industry, and out to harm the community, or that they knew these allegations were untrue but still drew on these as a means to stop my research. Our REB told me to go home, grieve the loss of my project and move on and pick another one. I was stressed and emotionally impacted by the feeling of both losing a great portion of my dissertation research but also feeling that people from my community, some of which I knew, had made personal attacks and allegations towards me.

Through complaints to our REB, local activists critiqued my dual role as community member and researcher, claiming I had a conflict of interest and motivations to further the facility proponents' agenda in an attempt to halt and presumably censor my research. The REB also initially criticized me for being 'too close' to the research, until one member reminded everyone that this 'closeness' was in fact encouraged in other contexts (e.g., Indigenous research, feminist

1 While this is currently Mason's doctoral dissertation research, she began this project as an MSc Candidate and then was accelerated into her doctoral degree.

studies) and should instead be seen as strength in my design. However, a portion of my research was eventually cancelled due to a procedural error, detailed below, uncovered following activists' complaints.

I was devastated when told I would have to discard my collected data and start over. What amplified the impacts of this event however was that my place of research and turmoil was also my home, normally a place of refuge. Where many students would have travelled home for the weekend to take a break, I returned 'home' immersed in the middle of conflict, accusations, confusion, disbelief, and media attention involving family, neighbors and friends – I could not escape. The situation took an emotional toll on my family while these community discussions continued for two months. The REB handled the issue by prioritizing the few complainants and hastily doing damage control as they rushed to review the file and make a final decision in an attempt to appease the activists' demands for a quick verdict (or they threatened to go to the media and pursue legal action) and mitigate further complaints. What they had not accounted for was that their quick actions with initially vague community messaging might negatively impact my family in the ways it did. This raises questions about the degree to which REBs account for impacts on immersed researchers (insiders or not) as well as the degree to which they consider the political motivations of activists trying to shut down research due to ostensible ethics violations. In the original letter sent to the complainants there was no explicit justification for termination. Instead, the following ambiguous statement was used to notify residents' of the termination of Mason's quantitative work in the original letter disseminated to Southgate activists: "After concerns were brought to the attention of University officials, an investigation was undertaken by the University's Office of Research Ethics, which in turn resulted in a decision to terminate the research study outright." This was left to residents' interpretation and resulted in the belief that it was their complaints (my involvement and supposed funding control by biased parties) that resulted in the cancellation of the research and not a separate procedural issue (described below) that was independently discovered. Further, the quantitative survey was mistakenly described as qualitative, resulting in confusion between the cancellations of the survey or interview portion of the research. The seven individuals who received this letter electronically circulated it throughout the community with their added interpretations. This and the REB's initial vague messaging eventually led to three community letters being disseminated with the last being a two page letter (mailed to all Southgate residents) explicitly clarifying each portion of the research, my involvement, and the procedural issue. Situations involving insiders perhaps deserve particular consideration, because the 'concern for welfare' we take such pains to protect for our research participants (Tri-Council, 2014), may be unintentionally damaging for insider researchers.

The series of events these prioritized complainants initiated with the REB were stressful, disheartening and certainly caught me off guard as a new researcher. Although I was encouraged to walk away from this research and do something 'easier', I still feel this is an important area of research, so the project continues while my family, community and I work on healing. There are many advantages to conducting research in your community, and I have had many positive experiences; however, I do caution qualitative researchers to consider the personal impacts this can have, regardless of REB's actions. As Buckle, et al. (2010, p. 119) warn, 'qualitative research is emotionally involving in ways seldom experienced by quantitative researchers', and I argue that this is even more so when conducting research in your own setting. This is not a reason to avoid 'insider' research, however an awareness of the potential impacts and emotional ramifications prior to conducting research should help prepare you for the backlash that can be faced, despite best efforts to the contrary.

Chad Walker

In some research there are more pressures to take sides that are increasingly difficult to negotiate, particularly when savvy activists are motivated to challenge our work. My master's research began as an investigation of daily life changes following wind energy development in southern Ontario. It was partially inspired by media reports of health problems, sleeplessness, and decreasing property values. In conversations with randomly selected residents of Port Burwell (a community with wind turbines) however, I found almost none of these to be occurring 'on the ground' – though several residents noted there were concerned and impacted people in nearby Clear Creek. After several weeks interviewing those supportive of wind energy, it was a shock to hear first-hand the problems some residents in Clear Creek were facing. Some would even break down and cry during our conversations. While I have become less surprised when these types of emotions are shared, I am no less disheartened when people feel an overwhelming anger or sadness regarding their particular living situation.

No matter whether a participant supports or opposes wind development, I feel the motivation for residents' participation in my research was about getting 'the truth' out there. Of course, one true reality does not exist, and this makes it difficult to sort through and represent residents' varied and often conflicting 'truths' or views of the same situation. While this chapter discusses my difficulties with some activists opposing wind turbines in their communities, proponents of wind have labelled me as being 'too sympathetic' to the concerns of those claiming health and other problems. Not fully supporting the convictions of either side of the debate meant I was left in a kind of purgatory – particularly regarding their trust.

The major question of my PhD research is focused on the causes and meaning of support (or opposition) of local wind turbines. It seems the question of causation is a point of contention for both sides. On one hand when I (among others) suggest that fair planning processes and financial arrangements increase support, those against wind turbines reply that realized impacts cannot be reduced so easily. Similarly, when I have reported perceived health problems and social conflict in wind turbine communities, proponents of wind dismissed the findings as insignificant. In a recent example, peer reviewers of a journal article suggested that only 34% of residents reporting property value loss due to turbines was 'not enough' to lead to a conclusion that turbines might result in financial loss. These examples suggest that the politics of wind turbines is pervasive and that trying to represent differing viewpoints is politically charged and problematic.

Introduction

What will you do when participants in your research make public claims to discredit or terminate your research? While this may not arise in much qualitative health geographic research meant to provide an empathetic perspective, such backlash lurks beneath the surface when studying environment and health controversies. Controversies such as toxic chemical remediation and compensation (Edelstein, 1988; Brown, 1994), facility siting for a proposed hazardous facility (Wakefield and Elliott, 2000) or policy change like a smoking or pesticide by-law (Hirsch, et al., 2010) tend to involve coalitions for and against the proposed action. Researchers often take an openly neutral stance on the subject matter in order to play a role gleaning insights from the situation for wider audiences. The stakeholders we involve in the research (e.g., in interviews or focus groups) may however be unsatisfied when the results are perceived to work against their stance on the issue – despite every effort we make to be just in the situation. Suspicions from stakeholders are likely to manifest when initial contact is made with potential participants, and may intensify as the research unfolds, especially when findings are shared which extend beyond the individual interview or focus group. This may be true when those we are calling activists – people who are heavily invested and publicly vocal about their particular stance on an issue – criticize our research. Increasingly, these activist publics are savvy to the inner workings of university systems and how these may be accessed to thwart the release of findings, particularly through the auspices of REBs. In our two case studies, such activists took direct actions to protect family, home and community against what they viewed to be serious threats in their communities. This chapter examines the increasing engagement of these community activists with REBs, alongside current critiques of REBs to shed light on the impacts activism and REB may be having on environment and health research.

Traditionally, discussions surrounding activism and academia in human geography have focused on enabling critical or radical geographies with the researcher's role as activist taking center stage (e.g., Castree, 2000; Blomey, 1994; Parr, 2004). For example, feminist researchers are concerned with both advocating for their female participants and exploring novel ways to encourage the empowerment of women as a whole (VanderPlaat, 1999). In this discussion of community activism and the changing nature of REBs, we are instead focusing on researchers' interactions *with* activists. Parr's (2010) discussion of research amidst legacies of conflict and handling disagreements over interpretations is one example, but papers like this are either few and far between or difficult to track down. That is, there is no literature that directly addresses issues of activism against academics, but some allied literatures include social movements and emotional aspects of research (Blee, 1998; Flam and King, 2007).

Yet, in terms of a major conduit of activism backlash, there is a growing literature that is critical of the role REBs[2] play as arbiters of social scientific research practice. They have been accused of being both overly restrictive (McCormack, et al., 2012; Dyer and Demeritt, 2008; Haggerty, 2004) and sidestepping some of the equally important issues of professional ethics, like working towards empowerment of marginalized groups (Murray, et al., 2012). Much of the ethics of research falls somewhere between what REBs can conceivably control and what various theories of research praxis suggest we might legitimately do. Further, the documents meant to guide ethical research conduct (e.g., Canada's Tri-Council Policy Statement, 2010) are open to interpretive inconsistencies surrounding REB principles such as 'do no harm' and 'informed consent' that are enacted at regulated institutions (McCormack, et al., 2012; Abbott and Grady, 2011; Angell, et al., 2006). REBs and universities fearful of the legal implications of their decisions have also been criticized for extending their mandate beyond the issues originally intended, including bias towards certain methodologies.

We draw on our own experiences to highlight difficulties that arise when the people who are participating in your research decide to rally against you and your study. This chapter will proceed first with a brief introduction of two environment and health case studies: health risk in communities living with (1) a proposed biosolid waste processing facility and (2) wind turbines in rural Ontario, Canada. These cases illuminate when activism against academic studies may become acute by contextualizing interactions between ourselves, the activists, and REBs. This will be followed by a more extended discussion of the role of REBs drawing on the notion of 'ethics creep', censorship and emergent designs in qualitative research. Further, we provide insights into how procedures meant to enhance qualitative rigor may be flashpoints for activist backlash, specifically member checking and autobiographies. We conclude with a discussion of lessons learned to suggest possible ways forward in environment and health and other potentially contentious research.

Urban biosolid processing in rural communities: facility siting, risk perceptions and community conflict

Sarah Mason is conducting her doctoral research within her community, the Township of Southgate in rural Southwestern Ontario, where she is using mixed methodologies to examine risk perceptions and social and emotional impacts of the siting and eventual operation of a regional biosolid (sewage sludge) to agricultural fertilizer processing facility. As the facility imports mostly urban biosolids into the small rural community, environmental justice concerns are among the many that ignited strong opposition within the community. Mason conducted semi-structured interviews with 23 adult Southgate residents in the summer of

2 Ethics boards are referred to as REBs in Canada, Institutional Review Boards (IRBs) in the United States and Research Ethics Committees (RECs) in the United Kingdom.

2012 in the middle of the contentious facility siting process (the facility became operational in 2013). Conflict within the community heightened as a local activist group opposing the facility carried out a three-month site blockade and took the municipality to court over land zoning issues.

Following initial analysis of the interviews, a survey was disseminated to all Southgate households in September 2012 looking to gain a broader understanding of residents' opinions before a final decision was made by the provincial environmental assessment panel – expected a month later. The survey immediately instigated actions by seven local activists who directly contacted the University REB. The complaints included Sarah's status as a community resident and former member of the facility's Public Advisory Committee and the belief that Mason and Luginaah (Mason's graduate supervisor and principal investigator of this research) were working for government and industry organizations to conceal negative facility impacts. The researchers were absolved of any ethical wrong-doing as far as the activist complaints were concerned, but following persistent complaints by these individuals the survey was cancelled due to a procedural issue uncovered by the REB itself, not the residents. Though the survey covered the exact same topics as the interviews, the specific survey questions were not added officially as an addendum to the original approval. It was maintained that the survey topic was no more harmful than the interviews, in the sense that no invasive questions were asked; however, REB representatives worried that the procedural issue (survey addendum) could arise if the activists had sued as they threatened to, and therefore the survey portion of research was halted.

In the short week that the REB deliberated upon the activists' complaints, 445 residents consented to, completed, and returned questionnaires prior to the first official announcement issued by REB, which stated the study was being terminated, but no explicit explanation was provided to the community.

The REB deemed the entire situation an 'adverse event' and further instructed Mason and Luginaah to cease research contact with the community and not publish results from the interviews. After persistent discussions with REB and Faculty representatives, along with requests for clarification with general community members, it was decided that materials from the interview transcripts could be published.

Worth noting is that the municipal government voted in favor of the facility itself, which became a strong rallying point as activists claimed these municipal officials failed to represent their broader constituents. Our survey research would have identified whether or not council had widespread support in the community (not the primary objective of our research but nonetheless a potential outcome). The irony here, given this is a book about qualitative methods, is that it was the survey that instigated the activist backlash, not the interviews. This was arguably because of timing – given the activists' final court appeal against the municipality, which challenged the zoning of the land the facility was proposed to be on (for a full description see Mason, et al., 2015), was denied coincidentally the week of survey distribution. This led some residents to believe the REB protest was the activists' 'last straw'. Had the interviews been conducted at this time, it is plausible they may have protested these as well. However, this duality may shed light on a broader belief that interviews can be seen as less threatening. While Mason's interviews captured the viewpoints of a handful of residents in depth, in the context of decision-making at the precise moment in the community, those interviews were less threatening than a method that could ostensibly measure public opinion community-wide.

Mason will soon be conducting follow up interviews with community members as the REB agrees enough time has passed (28 months) since the turmoil. Qualitative methodologies were selected not merely to avoid the upheaval previously experienced with the survey, but to gain a comparatively nuanced understanding of experiences and impacts following the onset of facility operations. Our return to the community is based on the notion of comparing pre- and post-facility siting views within the community.

Though we do accept our part in not submitting the proper addendum within the REB's procedures, we discuss the implications the REB's decision had on the ongoing research, the community, and Mason in her dual role as resident and insider researcher. The goal is to explore how research involving contentious issues and activism relate to a broad array of research issues including ethics, rigor, and the evolving nature and critiques of REBs.

Activist(s) against local wind turbines in Ontario

During his master's degree, Walker investigated community-level impacts and determinants of support for wind energy development in two southern Ontario communities. Wind turbines in rural Ontario have recently become politically divisive, particularly since enactment of the 2009 Green Energy Act, which was successful in streamlining renewable energy approval processes (Ministry of Energy, 2015) despite community objections (Songsore and Buzzelli, 2014).

In 2011 Walker conducted 26 in-depth interviews with local residents (24), and policy experts (2). This research uncovered majority support for wind turbines alongside significant changes to 'daily life' for some people (Walker, et al., 2014b), such as increased community conflict, perceived health effects (Walker, et al., 2015) and property value loss (Walker, et al., 2014a).

Interviews aimed to understand the in-depth 'daily life' experiences of living close to wind turbines in Ontario since no such case studies had yet been published. While Port Burwell was chosen as the original study site for being one of the earliest and largest implementer of wind turbines in the province, nearby Clear Creek was also chosen because Walker was hearing stories of their discontent from the Burwell residents. Clear Creek is also the community from which most of the criticism against our research originated. Harsh denunciations and subsequent threats by one individual to 'shut down' the study began shortly after the dissemination of our preliminary findings through member checking (discussed further below). Though Walker and Baxter received mostly positive replies from this process, two people known to oppose wind turbines had serious complaints with the data interpretations. Those complaints stated we did not go far enough to denounce the turbines that resulted in extended discussion with Walker and Baxter.

In terms of impact on our interpretation, we decided to focus the attention of our quotations in three papers on how the concerned citizens were impacted (Walker, et al., 2015; 2014a; 2014b) with far less coverage of the majority in support of turbines. This was meant to balance the academic coverage of turbine communities, which Aitken (2010) characterized as biased against concerned local residents. After member checking, there was a relative calm until the online publication of Walker's (2012) thesis. In the days that followed, we received emails from three individuals complaining about the thesis. Their biggest problem seemed to originate from the content of Walker's autobiography, which openly stated his biases going into and coming out of fieldwork (more detail below). Criticism spread from one participant across the province through wind turbine opposition websites with comments meant to discredit the research. The proclamations against our work were reignited each time a new academic publication emerged from our study. The claims against our study even included invocation of the Nuremberg Code concerning the unethical conduct of medical research on humans. That is, since we were reporting that the well-being of residents near turbines was negatively impacted (e.g., community conflict), we were tacitly supporting the ongoing and unethical exposure of those same people by doing the research.

The study was never shut down despite letters from the one participant to everyone from the departmental chair right up to the university's president. It is important to note these events occurred shortly after the REB's experience with activist engagement in Mason's research. Perhaps there were lessons learned by the University that led to caution and more thorough and participatory processes between the REB and Baxter and Walker than Mason and Luginaah had previously experienced in conceptually similar circumstances.

The role of the REB and reflections on the role of the activist

It was not that long ago that research ethics was simply a matter of professional integrity, but it has evolved into formalized procedures to protect potential research subjects and increase public trust and the accountability of experts. Basic principles of ethical research conduct have been articulated and reinforced including doing no harm, respect for persons, voluntary and informed consent, beneficence, and justice through a fair balance of risks and benefits (Brown, et al., 2010). Harmonized national principles for ethical research involving humans has only been in place in Canada since the 1998 publication of the first policy statement on ethics published by the three federal funding agencies (Tri-Council). Similarly formalized procedures have only been implemented more recently in much of the UK and EU. Despite, or perhaps because of, this relatively short history, there is already backlash against such formalized systems in Canada, in what we have already referred to as ethics creep (Haggerty, 2004; see also Dingwall, 2008). This creep refers to REBs being overly restrictive, delving into areas ostensibly beyond its mandate, and escalating both requirements and restrictions to the point where qualitative social scientists are focused too much on 'jumping through hoops' to achieve ethics approval that may ultimately do little to protect communities or researchers themselves (McCormack, et al., 2012). As different jurisdictions implement formalized procedures, reactions against ethics creep have emerged outside of Canada in such places as the United States (Becker, 2004) and the United Kingdom (Dyer and Demeritt, 2008), suggesting a systemic problem for social scientists.

In discussing 'ethics creep', Haggerty (2004, p. 392) describes how 'REBs have unintentionally expanded their mandate to include a host of groups and practices that were undoubtedly not anticipated in the original research ethics formulations'. This, he argues, suggests issues of 'institutional distrust' whereby it is presumed that researchers now require additional oversight through REB monitoring to ensure ethical decisions and actions are made. Further discussions regarding REBs overstepping their bounds have been identified as they move to assess the value, validity, rigor or practicality of social science research (Murray, et al., 2012; Dyer and Demeritt, 2008). For example, issues typically dealt with by thesis supervisory committees (e.g., methodology or sampling) have come under increasing scrutiny by REBs. Though friendly suggestions may always be welcomed in a system of checks and balances, Murray, et al. (2012) argues that REBs are in danger of absconding roles and in the process taking on responsibilities beyond their purview. Similarly, Dyer and Demeritt (2008, p. 3) suggest that REBs' 'wholesale and indiscriminate application will create more problems than it solves', whereby they take on the role of judge and jury on legitimate methods and methodologies.

As environment and health geography research often investigates communities in conflict and with publics increasingly skeptical of science, it is understandable that this skepticism also gets directed towards the ethical conduct of social science. Drawing upon Beck's (1992) risk society framework Haggerty (2004, p. 392) comments: 'Concerns about the ethical quality of research are characteristic of a society where anxieties . . . are increasingly common.' However, there may be a thin line between skepticism on the one hand, and censorship on the other hand when ethics are invoked ad hominem. While the awareness of rights and empowerment of activist participants with regards to institutional ethics boards should be applauded, REBs' fear of litigation should be weighed against intentional efforts to silence the voices of those with opposing views. In both case studies, activists attempted to stifle research they felt was not aligned with their viewpoints. In regards to Mason and Luginaah's survey research, residents spoke out through letters to the editor complaining against the cancellation of the research and the belief that activists were trying to censor the results – seen in the quote below from a resident who was neither a family member nor a close friend of Mason's:

> I, along with many others, am disgusted with the actions of a small group of residents who claim to know what is best for this municipality. . . . They hit a new

> low when they interrupted the education of a local Western University student claiming she had a 'conflict of interest'. Were they afraid the results of this survey would not support their propaganda? (*Dundalk Herald*, October 24, 2012).

In the case of the resident exposed to turbines, the REB decided there was no specific ethics violation. Though the board acknowledged that health and well-being were at least perceived as being impacted, the REB determined we as researchers were not the cause of the exposure. That said, we were mindful of how those individuals turning their sights on us, in some sense, meant we had become part of the 'exposure' – engaging in a conversation about turbines may have been bringing forward negative emotions. Subsequently we repeatedly suggested the one particularly determined participant consider withdrawing from the study. In the end, without confirmation that the person wanted to remain in the study, we withdrew them. It is somewhat ironic that we (including the REB) felt that withdrawal would remove this person's access to the very structures in the University he seemed to want to hear his case (i.e., the REB, deans and the president). In such situations it is important for REBs to consider the merits of the ethical claims of concerned citizens and determine whether the claims are tools that can be used by activist publics to forward their interests.

Murray, et al. (2012, p. 46) suggest a more balanced motivation for REBs whereby they ask, 'How can we facilitate this research and not just block it?' As researchers, we agree that research can be increasingly controversial when conducted in contentious communities; however, as Canada's Tri-Council (2014, p. 20) policy states, REBs 'should not reject proposals because they are controversial, challenge mainstream thought, or offend powerful or vocal interest groups'. With the necessity to obtain REB approval to conduct social science research, REBs also have the power to determine how human geography research is conducted and subsequently what environment and health problems are investigated. From these experiences, we have learned that by prioritizing an individual participant without substantive claims, the ethics review processes fail to support critical scholarship seeking to expose injustices, academic freedom to inquiry, and the public right to know (Dyer and Demeritt, 2008).

Anticipation of harm in inductive research and procedural ethics

The notion of emergent designs in qualitative research is particularly relevant to health research where changes may happen rapidly on the ground. As we enter communities seeking to examine an environment and health phenomena, often-unforeseen themes emerge giving way to additional questions or conceptually important participant categories. This 'messy' approach is how nuanced findings and deeper understandings of contextual experiences are theorized (Denzin, 1997). Yet, it is these investigative uncertainties that create the greatest 'ethical quandaries' for researchers (McCormack, et al., 2012, p. 33). It is impractical for REBs to demand anticipatory prescriptive project plans for exploratory qualitative research and subsequently penalize researchers when the plan shifts (e.g., new participant groups, larger sample sizes). Pollock (2012) distinguishes between institutional procedures as rules within REBs versus in the field processes as defined by professional ethical conduct (see also Guillemin and Gillam, 2004) and argues that at many levels ethical decisions would be best monitored through 'micro' or process ethics based on judgements rather than a priori application of rules. We do however recognize that a 'judgement' based process would only introduce more subjectivity and interpretive power to members of REBs, which depending on methodological biases among committee members may or may not be helpful to decisions regarding nuanced qualitative research. Though 'deviations' can be brought to a REB's attention (i.e., amendments forms) such a process can work against happenings 'in the field'. As REBs collect particular details such as sample size, there is concern that the information collected on the application represents a contract of sorts. An increasing awareness of the willingness of REBs to view their work in legalistic, contractual terms opens up research further to stoppage by shrewd activists. In our view, REBs should consider the merits of emergent qualitative methodologies on a case-by-case basis, thereby considering issues of broader moral ethics. Haggerty (2004) argues the pendulum has swung too far – towards a rule-based system rather than one based on core ethical principles:

> [With] the fetishization of rules . . . researchers risk being seen as acting unethically when they fail to submit an application to the REB or to obtain a signed consent form, whether or not there was ever the slightest prospect of anyone being harmed by virtue of such research (Haggerty, 2004, p. 410)

New ethical tensions created by the constantly evolving nature of qualitative health research 'in the field' and the propensity for REBs to become enforcers of 'rules', has shifted the original intent of the REB to ensure principles of 'do no harm' and informed consent. Further, Haggerty (2004) reinforces the injustice of such procedural decisions, saying it is 'divorced from common sense' (p. 411). This fracture bore out in Mason's research when such a procedural issue resulted in cancellation and silenced 445 consenting individuals.

Qualitative rigor

REBs are only one aspect where community activists are involved in research. In fact, hostile responses to our work may be due more so to our own well-intentioned research decisions, such as engaging participant feedback on the data interpretations. In the case of wind turbine work, Walker engaged two procedures intended to strengthen the qualitative rigor (credibility) of the study: member checking and autobiography (Baxter and Eyles, 1997; Lincoln and Guba, 1985). While these procedures were used to increase the transparency of the research, in combination these strategies fuelled the flames of discontent, particularly with two primary objectors.

Member checking

Member checking is used to grant research participants access to (initial) data interpretations while establishing the relationship between the researcher's and participants' perspectives of those interpretations (Sandelowski, 1993). This approach aids the iterative process of interpretation, helping to situate the data within the context of the actor's 'true' point of view (Bloor, 1983; Hoffart, 1991). Indeed it may be informally done during each interview when an investigator asks for clarifications (Sandelowski, 1993), but our focus here is on a more formalized process later in the interpretive phase of fieldwork.

There is some debate about the value of member checking since it can make fairly opaque interpretive processes more frustrating, while others worry about loss of interpretive sovereignty. However, Lincoln and Guba (1985, p. 314) suggest that member checking is 'the most crucial technique for establishing [research] credibility'. Yet, Sandelowski (1993) warns that member checking may present more problems than it solves and may actually undermine the trustworthiness of a study if it fails to recognize the diverse interpretative goals of the researcher(s) and participants (see also Turner and Coen, 2008; and Hammersley, 1992). In this context, Walker's primary analytic goal was to increase 'richness' of the data through further interpretative input and reflection and not achieve participant consensus. To achieve this goal, participants were mailed a document that briefly outlined preliminary results along with an explanation of the member checking process. This document highlighted the nuances of support and opposition to wind turbines in Port Burwell, among other things. Participants were asked if they agreed with the document and requested to comment extensively. Of the 12 who responded, most (8) agreed with the findings, and two participants strongly disagreed. The two participants with opposing views also vehemently opposed the turbines, and in addition one of these two participants provided an extensively written (i.e., 4,100 word) email (followed with many others) in response. It is this particular response that Walker found difficult to reconcile against the other findings, which pointed to high levels of support.

A common theme throughout this email response was the accusation that our findings were 'missing the mark' and the claim of misrepresenting the actual situation in Clear Creek. Perhaps because an inherit value of qualitative work is the ability to represent marginal, disadvantaged populations rendered 'inconsequential' by other (quantitative) methodologies (Watters, 1989), this person felt we did not present their views and their situation (e.g., health impacted) on the same level as those supportive or accepting of wind turbines. The opponents' quotes made up only about 27% of the document while a follow-up survey later revealed only 23% of people in both communities opposed their respective community wind projects. That said, the objections gave us pause to consider focusing more attention on the concerned citizens to support a greater voice for their issues in the academic literature and ultimately the local media.

An additional objector accused us of biasing our participant selection – that Port Burwell residents were selected because they were all lease holders and were financially benefitting from turbine development. Though this was not the case – only two residents (of 26) with a turbine lease were interviewed – this situation did remind us of the high stakes: that our overall design was under scrutiny. This member checking response letter, from another participant than discussed previously, also highlights the issues that member checking may open up the research up to more personal attacks. This resident wrote that the work was too full of 'flotsam and jetsam' with 'no valid conclusion', further suggesting it was comparable to 'that of a high school student'.

While Mason had not yet conducted member checking within her research community, following the turmoil that arose from the contested survey, she was advised to avoid returning to the community to conduct member checking and examine interview interpretations. It was the REBs concern that this process would ignite further opposition by the activist community – understandable given Walker's experience. However, we do note that REBs' aversion to community conflict does have implications for qualitative rigor within research.

Autobiography and revealing yourself

Autobiographies are recommended to augment the process of reflexivity or self-awareness of bias during the research process, thus Walker included one in his MA thesis. This was subsequently scrutinized by one concerned citizen research participant and, based on his reference to the autobiography in further communications, seemed to help precipitate this person's claims that we were not conducting ethical research. This autobiography required Walker to not only disclose biases going into the field, but also reveal how the process of qualitative research changed those biases and ultimately Walker's identity. Autobiographies are intended as a corrective to positivistic research, which assumes researchers can enter and exit the field as objective automatons; its intent is in the spirit of openness to enable the reader to better understand the context and motivations for the research.

Determining what exactly to include in the autobiography was a difficult decision, partly because it is rarely done. It is frequently used as an independent research method in itself – widely encouraged in the social sciences disciplines such as anthropology (Okely and Callaway, 1992), education (Mitchell, et al., 2013), sociology (Harrison and Lyon, 1993) and any discipline that incorporates a self-reflecting approach to ethnography (Hannabuss, 2000). Walker's autobiography included general comments about his education and research interests:

> My interest in the current research on wind energy and more generally renewable energies has developed from a love of good environmental stewardship. . . . [During my undergraduate years] I found there were apparent disconnects between climate change science and policy and enjoyed studying the many proposed solutions designed to marry the two (Walker, 2012)

However, in the spirit of the theoretical use of autobiography, Walker also revealed initial feelings on core research issues he later learned he would be forced to reflect upon throughout the thesis work and beyond, particularly the issue of wind turbine-induced health effects:

> I became much more sympathetic to the problems facing the people I spoke with who are facing difficulties. . . . [However] I feel that any type of problem created with the introduction of wind turbines into a particular area appear to be smaller in comparison with the ecological, human health, and social impacts associated with a continued reliance on GHG-rich sources of electricity. (Walker, 2012)

In the days after Walker's thesis was published, the autobiography prompted response from one particular participant who argued that Walker's bias disclosure meant the entire work should be dismissed. In terms of the ethics process, it was the autobiography combined with member checking that prompted what this person felt were legitimate claims against the entire study.

Activism in the media and online

Both projects were scrutinized online, similar to other studies and papers in such controversial fields of study. For months after Walker's thesis publication, there was a relative absence of activism against the research until early 2014 when journal articles started to be published. These articles received significant media attention. Though we maintain that the papers are actually far more sympathetic to concerned citizens than the majority of academic writing on turbine communities to date, we were obligated to point out that the majority of people we interviewed and surveyed in all cases were supportive of turbines. However, it is most telling that Walker and Baxter have published papers highlighting the health impacts of turbines from the point of view of conflict, community ostracization, and well-being. The majority of online comments were associated with the first paper we published (Walker, et al., 2014b) and are linked to an article in a local paper stemming from our own press release. A total of 138 mostly negative comments have since accumulated (Miner, 2014). These comments vary, but the main theme is that the research was biased and speculated to be funded by industry or the Liberal government. Similar assumptions and accusations were made towards Mason's research through social, online and print media. Yet, in both cases, we clearly acknowledged the university as the sole source of funding.

The efforts to discredit the research highlight that wind turbine and waste processing research is clearly politically divisive in Ontario, but more insidiously it spotlights the pressure on researchers to yield to one point of view or another. This is likely not the context that ethnographers had in mind when warning each other of the lure of 'going native' by losing touch with academic principles and audiences in the service of the people we study, but the same basic principle applies.

Developing a 'thick skin' is an understatement in this type of work when your research goes on trial publicly to be characterized dubiously as an 'absolute disgrace. . . . My 12 year old son could do a better, more accurate job. My mark for this work is F.' There were also other mentions of Walker's potential bias found through other online forums such as the accusation that his positive comments about wind energy as an undergraduate student in the BG News (Ohio, USA) in 2009 created an implied bias three years later. Mason and Luginaah experienced similar backlash where activists 'dug deep' using social media, news sources, and career associations to suggest similar historical bias.

Implications and lessons learned

Our experiences with activists and REBs have sensitized us to what may be a new era of research for social scientists interested in controversial topics. Not only are activists generally knowledgeable, savvy and increasingly engaged, they are resourceful and willing to go to great lengths to use institutionalized and rule-oriented REBs. We should be heartened somewhat, and ironically, that our desire for empowered publics is being realized. There is a not-so-fine line though between accusations of unethical research practice and outright efforts on the part of groups within communities at censorship. This should serve as a reminder to researchers to be well prepared up front and that satisfying REB requirements is a necessary but not sufficient preparation for handling activists concerns.

Further, REB approval is insufficient for wrestling with deeper ethical issues such as finding ways to represent those same activists, who are often marginalized in their local and wider communities. We still feel it is our role to help make their voices heard in empathetic ways despite what they may say against us as researchers. It is indeed challenging to juxtapose hegemonic viewpoints (e.g., biosolids and turbines are minimal health risk) against these marginalized voices, since research generally tends to favor majorities. Fortunately, with qualitative research we have the opportunity to more fully represent the marginalized views that strictly quantitative designs typically understate. The delicacy lies in the fact that representing alternative opinions can instigate criticism from both academia and majority publics who, for example, prioritize rural sustainability, action on climate change, or pollution reduction.

Given the inherent challenges, a reasonable alternative to tacitly claiming objective impartiality is to focus efforts on transparency, but this comes at a cost. Regardless, it is essential to monitor and reflect on our own positionalities, either openly in our publications or behind the scenes. It is a tough position to take the risk of exposing yourself to criticisms and possibly censorship while simply being honest about bias in an effort to contain it. This will make the task of multidisciplinary REBs more challenging. When research becomes politicized by community activists claiming ethics violations, REB members with little experience of qualitative methods may find it difficult separating ontological and epistemological differences within their own biases with matters of ethical principal. For example, efforts towards improving rigor through transparency are a legitimate component of much qualitative and critical inquiry and not (necessarily) a signal that ethics have been breached by incompetence to conduct sound (read: 'unbiased') research.

Further, it is a difficult decision to resist 'setting the record straight' as your personal and professional reputations are attacked. Our experience is that outside the context of an interview the exchange is largely one-way and political – though perhaps, understandable given these voices have been ignored in many other contexts. When considering this we have had two thoughts: (1) silence makes a tacit statement that we stand by our work, and (2) fighting back takes effort that could otherwise be put into thoughtful research and may merely act to instigate further assault. Yet, in line with Haggerty's (2004) notion of ethics creep and increasingly legalistic nature of REBs (Pollock, 2012), important research may be at risk of being shut down or never seeing the light of day for fear of political repercussions.

We are not suggesting to eliminate REB review, however as this is an increasingly contested institution that likely disproportionately impacts qualitative researchers (Pollock 2012; Dyer and Demeritt, 2008), we feel it useful to weigh in concerning what may be a new era of activist backlash. We simply ask that REBs consider the theories behind the methodologies and act in proportion to the associated risks. Further, while we do not feel that qualitative research is directly under attack by REBs, we remind that such attacks tend to be incremental and perhaps even unconscious, yet the end result is that dismantling qualitative research of any sort is another form of censorship. We urge REBs to take a step back and focus on core principles such as balancing benefit against harm, respect for persons, beneficence and justice. Ultimately the rule-based procedural ethics of REBs must give way to professional ethics in the field where researchers are tasked with making ethical decisions on a regular basis. Even when we do act ethically according to professional and disciplinary standards, that will do little to prevent community activists – marginalized or not – from taking our research to task.

References

Abbott, L., and Grady, C., 2011. A systematic review of the empirical literature evaluating IRBs: what we know and what we still need to learn. *Journal of Empirical Research on Human Research Ethics*, 6(1), pp. 3–20.

Aitken, M., 2010. Why we still don't understand the social aspects of wind power: a critique of key assumptions within the literature. *Energy Policy*, 38(4), pp. 1834–1841.

Angell, E., Sutton, A., Windridge, K., and Dixon-Woods, M., 2006. Consistency in decision making by research ethics committees: a controlled comparison. *Journal of Medical Ethics*, 32(11), pp. 662–664.

Baxter, J., and Eyles, J., 1997. Evaluating qualitative research in social geography: establishing 'rigour' in interview analysis. *Transactions of the Institute of British Geographers*, 22(4), pp. 505–525.

Beck, U., 1992. *Risk society: towards a new modernity*. London: Sage Publications.

Becker, H. S., 2004. Comment on Kevin D. Haggerty, 'Ethics creep: governing social science research in the name of ethics'. *Qualitative Sociology*, 27(4), pp. 415–416.

Blee, K. M., 1998. White-Knuckle Research: Emotional Dynamics in Fieldwork with Racist Activists. *Qualitative Sociology*, 21(4), pp. 381–399.

Blomey, N., 1994. Activism and the academy. *Environment and Planning D, Society and Space*, 12, 383–385.

Bloor, M., 1983. Notes on member validation. *Contemporary field research: a collection of readings*. Boulder, CO: Westview Press. pp. 156–172.

Brown, P., Morello-Frosch, R., Broday, J. G., Altman, R. G., Rudel, R. A., Senier, L., Perez, C., and Simpson, R., 2010. Institutional review board challenges related to community-based participatory research on human exposure to environmental toxins: a case study. *Environmental Health*, 9, pp. 39–50.

Brown, P., and Masterson-Allen, S., 1994. The toxic waste movement: a new type of activism. *Society and Natural Resources*, 7(3), pp. 269–287.

Buckle, J. L., Dwyer, S. C., and Jackson, M., 2010. Qualitative bereavement research: incongruity between the perspectives of participants and research ethics boards. *International Journal of Social Research Methodology*, 13(2), pp. 111–125.

Castree, N., 2000. Professionalization, activism and the university: whither 'critical geography'? *Environment and Planning A*, 32, pp. 955–970.

Denzin, N. K., 1997. *Interpretive ethnography: ethnographic practices for the 21st century*. Thousand Oaks, CA: Sage Publications.

Dingwall, R., 2008. Thc ethical case against ethical regulation in humanities and social science research. *21st Century Society*, 3(1), pp. 1–12.

Dyer, S., and Demeritt, D., 2008. Un-ethical review? Why it is wrong to apply the medical model of research governance to human geography. *Progress in Human Geography*. doi:10.1177/0309132508090475.

Edelstein, M. R., 1988. *Contaminated communities: the social and psychological impacts of residential toxic exposure*. Boulder, CO, US: Westview Press.

Flam, H., and King, D., 2007. *Emotions and social movements*. London and New York: Routledge.

Guillemin, M., and Gillam, L., 2004. Ethics, reflexivity, and 'ethically important moments' in research. *Qualitative Inquiry*, 10(2), pp. 261–280.

Haggerty, K. D., 2004. Ethics creep: governing social science research in the name of ethics. *Qualitative Sociology*, 27(4), pp. 391–414.

Hammersley, M., 1992. Some reflections on ethnography and validity 1. *Qualitative studies in education*, 5(3), pp. 195–203.

Hannabuss, S., 2000. Being there: ethnographic research and autobiography. *Library Management*, 21(2), pp. 99–107.

Harrison, B., and Lyon, E. S., 1993. A note on ethical issues in the use of autobiography in sociological research. *Sociology*, pp. 101–109.

Hirsch, R., Baxter, J., Brown, C., 2010. The importance of skillful community leaders: understanding municipal pesticide policy change in Calgary and Halifax. *Journal of Environmental Planning and Management*, 53(6), pp. 743–757

Hoffart, N., 1991. A member check procedure to enhance rigor in naturalistic research. *Western Journal of Nursing Research*, 13(4), pp. 522–534.

Lincoln, Y., and Guba, G. (eds.), 1985. *Naturalistic inquiry*. Newburry Park: Sage.

Mason, S. A., Dixon, J., Mambulu, F., Rishworth, A., Mkandawire, P., and Luginaah, I., 2015. Management challenges of urban biosolids: narratives around facility siting in rural Ontario. *Environmental Planning and Management*, 58(8), pp. 1363–1383. doi: 10.1080/09640568.2014.925853.

McCormack, D., Carr, T., McCloskey, R., Keeping-Burke, L., Furlong, K. E., and Doucet, S., 2012. Getting through ethics: the fit between research ethics board assessments and qualitative research. *Journal of Empirical Research on Human Research Ethics*, 7(5), pp. 30–36.

Miner, J., 2014. Western University researchers calling on governments and wind farm developers to avoid feeding war of words. *London Free Press*. May 21, 2014. http://www.lfpress.com/2014/05/21/western-university-researchers-calling-on-governments-and-wind-farm-developers-to-avoid-feeding-war-of-words.

Ministry of Energy, 2015. Renewable Energy Development: A Guide for Municipalities – Updated for FIT 2.0. Renewable Energy Facilitation Office; Ontario Ministry of Energy. http://www.energy.gov.on.ca/en/files/2016/01/municipal_guide_english_web_2016.pdf. (Accessed January 29, 2016).

Mitchell, C., O'Reilly-Scanlon, K., and Weber, S. (eds.), 2013. *Just who do we think we are? Methodologies for autobiography and self-study in education*. Abingdon, Oxon: Routledge.

Murray, L., Pushor, D., and Renihan, P., 2012. Reflections on the ethics-approval process. *Qualitative Inquiry*, 18(1), pp. 43–54.

Okely, J., and Callaway, H. (eds.), 1992. *Anthropology and autobiography*. Psychology Press.

Ontario Wind Resistance (OWR), 2013. Responses to 'Winds of change may be blowing for wind turbines, study suggests.' July 24. http://ontario-wind-resistance.org/2013/07/24/winds-of-change-may-be-blowing-for-wind-turbines-study-suggests/. (Accessed January 29, 2016.)

Parr, H., 2004. Medical geography: critical medical and health geography? *Progress in Human Geography*, 28(2), pp. 246–257.

Parr, J., 2010. 'Don't speak for me': practicing oral history amidst the legacies of conflict. *Journal of the Canadian Historical Association*, 21(1), pp. 1–11.

Pollock, K., 2012. Procedure versus process: ethical paradigms and the conduct of qualitative research. *BMC Medical Ethics*, 2012(13), pp. 25–36.

Sandelowski, M., 1993. Rigor or rigor mortis: the problem of rigor in qualitative research revisited. *Advances in nursing science*, 16 (2), pp. 1–8.

Songsore, E., and Buzzelli, M., 2014. Social responses to wind energy development in Ontario: the influence of health risk perceptions and associated concerns. *Energy Policy*, 69, pp. 285–296.

Tri-Council Policy Statement, 2014. Ethical conduct for research involving humans: interagency secretariat on research ethics. Ottawa, Ontario, Canada: Public Works and Government Services.

Turner, S., and Coen, S. E., 2008. Member checking in human geography: interpreting divergent understandings of performativity in a student space. *Area*, 40(2), pp. 184–193.

VanderPlaat, M., 1999. Locating the feminist scholar: relational empowerment and social activism. *Qualitative Health Research*, 9(6), pp. 773–785.

Wakefield, S., Elliot, S. J., 2000. Environmental risk perception and well-being: effects of the landfill siting process in two Southern Ontario communities. *Social Science and Medicine*, 50, pp. 1139–1154.

Walker, C. J., 2012. 'Winds of change': explaining support for wind energy developments in Ontario, Canada. Master's thesis, University of Western Ontario.

Walker, C., Baxter, J., Mason, S., Luginaah, I., and Ouellette, D., 2014a. Wind energy development and perceived real estate values in Ontario, Canada. *AIMS Energy*, 2(4), pp. 424–442.

Walker, C., Baxter, J., and Ouellette, D., 2014b. Beyond rhetoric to understanding determinants of wind turbine support and conflict in two Ontario, Canada, communities. *Environment and Planning A*, 46(3), pp. 730–745.

Walker, C., Baxter, J., and Ouellette, D., 2015. Adding insult to injury: the development of psychosocial stress in Ontario wind turbine communities. *Social Science and Medicine*, 133, pp. 358–365.

Watters, J. K., and Biernacki, P., 1989. Targeted sampling: options for the study of hidden populations. *Social problems*, 36(4), 416–430.

Part II

Representation, self and community

5 Writing illness through feminist autobiographical analysis

Pamela Moss

Reflections on praxis

Writing illness for me has been a way to thread my interests in bodies, power and space with feminism and feminist methodology. I find that bodies are a particularly rich site for understanding how a relational notion of power (after Foucault, 1990) generates the self and the subject over and over again in various contexts that are just a little bit different each and every time depending on the context within which the bodies live. The specificity of each moment, where *a* self materializes and *a* subject takes form, fascinates me. I organize my research to figure out *how* this happens. Writing illness is one way to put into words the curious, thought-provoking, and noteworthy aspects of the tracings of this process that I have observed in my research. I draw on feminist understandings and interpretations of the content of the tracings to see the effects of these materializations so that I can place illness in relation to wider systems and structures in hopes of effecting social change.

I have engaged in writing illness in different ways. I have written illness as singular individual stories of the breakdown of physical bodies (e.g., Moss, 1999; 2008; 2014b), as segues into a discussion of social interconnectedness in the negotiation of space for older women living with chronic illness (Moss, 1997), and as a composite of several women's lives in order to emphasize wider themes I found in a large research project (see chapter epigraphs in Moss and Dyck, 2002). I have also written illness beyond the direct experiences of individuals by relying on historical documents to trace how individual soldiers receive care on the battlefield (Moss, 2014a). I find writing illness in these ways rewarding, for they yield insight into the discursive and material effects of the exercise of power in, on and through bodies in particular places.

Yet the most challenging way of writing illness for me has been writing my *own* illness. After having received my first large research grant for a project investigating the lives of women who had been diagnosed with either rheumatoid arthritis or myalgic encephalomyelitis (ME), I got sick. And then I got sicker. And then I was diagnosed with ME. And then I spent a year and a half in bed, trying to rest. As I lay there, thinking about my body, my job, and the people in my life, I began fashioning ways to manage my body so that I would be able to work and maintain relationships with people who were important to me. Health-wise, I cared for myself through diet, rest, and homeopathy. Personally, I welcomed household and emotional support from friends and family who accepted me as a person with ongoing illness. Intellectually, I directed my efforts toward reading about autobiography and using autobiographical writing as a way to theorize bodies, power, and space through the fatigue and pain I live with every day. I work from the premises that my life is not unique and that my writing is neither therapeutic nor intensely self-absorbed; rather, I use the method of writing my own illness as a way to draw out how power gets exercised in the multiple realms I traverse professionally, medically, and institutionally.

Once I got more deeply immersed in autobiography, I saw that as an analytical method, autobiographical writing could assist me in addressing some of the other research interests I had in feminist methodology, such as, knowledge production, critical reflexivity, and relationality. Theoretically, it made sense for me to query subject formation and to try linking the volatility of illness with identity. These pursuits led me to re-engage with writing illness to think about what ill bodies *do* – not in the sense of physical restrictions on or social limitations of the body as some narratives might, but in the sense of facilitating processes that generate effects, sometimes in generative ways, sometimes in obstructive ways. For example, ill bodies can draw attention to how employment policies are overly reliant on categories that define bodies as either fully abled or fully disabled, without an in-between to show the need for more embodied policies (Moss, 2000). Ill bodies can also be frontline fodder within the neoliberalizing academy with the intensification of labor (where managed rest is not an option), streamlined organizational restructurings (where decisions about specific bodies can be folded into wider system reorganizations), and cuts based on financial exigency (where accommodation plans are reconfigured by cutting access to physical supports and resourcing) (Moss, 2012).

For as long as I can remember, my ill body has informed my academic writing. Yet most of my published academic work has none of the sticky residue of my unrelenting pain and debilitating fatigue, which somehow vanish through the process of revising and rewriting. The pieces that do keep the gummy surface do not ever really capture what my life is like, for they too have been transformed into something publishable through peer review and more revision. I still have several drafts of papers that use autobiographical writing as feminist analysis as a way to write my illness – some of which are about subjectivity, some about feminist politics, and some about research methods. As a researcher, I know that they will remain unpublished because they cannot be written academically without linking the micro-scale of experience to wider processes. I understand my work to be at the margins of critical health geography and part of a theoretical tradition that values analyses of productive power, generative ontologies, and an affirmative politics.

Writing illness in qualitative health geography

In health geography, writing illness is a common way to humanize disease, to put a face to processes that are not always visible. Health geography researchers commonly distinguish between disease as a biological or physiological process and illness as the lived experience of disease. Whether life-threatening, life-altering, or something that has yet to come to fruition, the experience of disease is something worth writing about (e.g., Angus, 2008; Davidson and Henderson, 2010; Thien and Del Casino, 2012). For qualitative health researchers, asking people about their experience of disease is the foundation of many scholarly research programs. These types of studies show that individuals negotiate the breakdown of bodily systems in very specific ways, whether they be acute episodes or chronic conditions (e.g., Moss, 2003; Sultana, 2012; Donovan, 2014). Querying individual experience complements information collected through clinical observation and medical testing, which form standard protocols for assessing health and treating disease (e.g., Crooks, et al., 2011; Liaschenko, et al., 2011; Liggins, et al., 2013). Qualitative research about illness assists policymakers, physicians, and bureaucrats in health ministries and insurance companies in shaping healthcare in the provision of treatment options and what is effective in the delivery of health services (e.g., Castleden, et al., 2010; Curtis, 2010).

Coming from my particular interests in writing illness, I have to ask myself: what of people's own stories of illness? What do they have to say about the things health geographers are interested in? Given that health and ill bodies are usually under the purview of medicine and medical practitioners, much of the qualitative health research in geography is oriented toward engaging with the medical system in some way, even when the research is a critique of medicine. Given that most health research is grounded in the same scientific knowledge base as both biomedical research and medical practice, writing one's own illness does not usually count as much as when a researcher reports someone else's experience of disease. Personally written stories are marginalized, deemed as idiosyncratic and ignored as having little to contribute to wider discussion of health and disease. As well, within science, the body is usually treated as an object that can be studied – prodded, poked, punctured – and that can be known through anatomy, physiology, cellular biology and neurology, among other systems and structures. Thus, within medicine, experience of illness is restricted to consultative visits between a physician and the patient. Through these encounters, which can range from long-standing relationships to fleeting moments between strangers, bodily sensations reported by the patient get translated into medical parlance via designating some of these sensations as symptoms, for we know that doctors' visits shape the way disease is conceptualized and illness is experienced (Crooks, et al., 2012; Reiser, 2009).

Yet the steady stream of individual stories, written by individuals themselves, located on the edges of the health geography literature, scattered across sociology, anthropology, and disability studies, shows that writing illness must have something to say about illness beyond the mere provision of idiosyncratic accounts of being ill, living with illness or having an ill body. Writing one's own illness rests on a set of competing assumptions about what can count as a way to know health and know an ill body: experience is valued as a way of knowing the body. In writing illness, experience and emotion are both meaningful ways of knowing and form a basis from which knowledge about the body can emerge. Understanding that autobiography can be an effective research approach to access salient information about various experiences and emotions of ill bodies reframes how qualitative health research can be undertaken. Autobiographical stories of bodies with broken systems can inform answers to the questions health geography researchers are interested in, including subject formation, political practices, and everyday life.

In the rest of this chapter, I first discuss two key elements of writing illness – experience and illness – in light of autobiographical writing as an approach to feminist analysis. I next locate feminist analysis and autobiography within and outside health geography and provide an overview of the various ways feminist researchers in geography have used autobiographical writing as part of understanding experiences of health and illness. I then provide a conceptual framing I have found useful in understanding my own use of feminist autobiographical writing. As a way to illustrate these arguments, I offer an analysis of my own illness. I close with a set of remarks about the effectiveness of writing illness through feminist autobiographical analysis.

Experience and illness through feminism and autobiography

There is a long-standing affinity between feminist methodologies and valuing experience as a credible source of knowledge in reading and writing autobiography (Smith and Watson, 1998). Autobiographical writing is a type of feminist analysis that brings out the specificity of experience without ignoring the wider social and cultural processes of being ill. As an approach to research, autobiographical writing can enhance understandings of the ways in which individuals negotiate meanings of health and illness within and outside medical notions of the body, identify key pieces of information used to make decisions about medical treatment and act as exploratory research to find innovative sites of inquiry. For those already part of critical research, by engaging in thoughtful reflection and accounting for intersubjective practice, researchers can generate rigorous, theoretically informed analyses of one's own experience of illness that can contribute to understanding wider questions about the body, power, and space in health geography.

Legacies

Central to writing illness is the concept of experience. Two pieces in the sociological literature have heavily influenced the way researchers in the social sciences, including geography, understand the experience of illness. The first piece, key in popularizing the social model of disability in disability studies, introduced the idea of 'biographical disruption' in relation to being ill with chronic illness. Bury (1982) interviewed 25 women and 5 men with rheumatoid arthritis. He purposefully chose people who were in the initial stages of the illness so as to highlight the ways in which people make adjustments in their lives. The second piece emphasized the importance of dealing with, rather than succumbing to, severe, life-threatening illness. Frank (1991) wrote about his reactions emotionally and physically to both illnesses he endured, heart attack and testicular cancer. He drew out some of the problematic aspects of living with illness in a society that values health that he encountered through interaction with friends, acquaintances, and medical practitioners as well as the wider conventional, societal understandings of illness. He conceptualized illness as living through the disease.

Both Bury and Frank attempted to capture the *experience* of a disease, what it is like to *be* ill. Bury reinforces the idea that chronic illness is a disturbance to the trajectory of one's life, some kind of commotion that needs to be managed. Frank, too, stresses the disruption to one's life. The idea of illness as disruption casts those with illness as not being able to live the life planned and subsequently restructures one's life path as somehow second-rate to the way one would want to live. Despite recognizing that medical treatment and social networks influence the way people in the lives of those with illness interact with the person who is ill, both scholars still tie the notion of the experience of illness to the body of the individual: they locate the experience of illness at the juncture of the deterioration of the physical body and the understanding of one's self. Thus, illness – both acute and chronic – acts as a derailment of sorts and becomes something to fight against in a refusal to let illness define one's identity.

Feminism and autobiography

One of the questions that feminists were dealing with at the time these works were written was what does experience actually comprise? Feminists interested in autobiography took up the question of experience in relation to questions about truth, difference, power, subjectivity, and identity (see, e.g., Brodzki and Schenck, 1988; Miller, 1991; Gilmore, 1994; Coslett, et al., 2000). These works took experience as something other than a universal condition of being human. Through bracketing, a researcher or an author could peel away the layers of society and culture and expose an essential structure of consciousness. Feminists challenged this humanist conceptualization of experience – a notion that both Bury and Frank rely upon in their analyses – by offering a more contingent notion of experience. For them, experience itself was constituted by the sets of relations giving rise to, in this instance, the encounter of illness (see Scott, 1991). A humanist understanding of experience might develop a line of thinking that tries to access the essence of illness. By focusing on stripping off the compacted layers of societal and cultural meaning, humanist analysts strive to keep separate the reactions to the individual's experiences of illness by family, friends, and acquaintances, and the assessments by physician's, surgeons and specialists so as to access an individual's experience of illness. Feminists rethinking experience might claim that illness itself is the materialization of multiple factors that together come to form an experience of illness. For these feminists, experience of illness is a bit of a shemozzle that only takes an ordered form *through* encounters with family, friends and acquaintances as well as physicians, surgeons and specialists.

The orderliness that makes illness tangible does not fix experience of illness as one thing; rather, the orderliness loosens with the next encounter and the experience of illness changes just a little bit. Encounters over time continually shape and reshape the way individuals live with illness and their ill bodies as well as how they come to understand themselves as ill.

The idea that illness is something arising out of the very situation individuals find themselves in, as part of a set of encounters dependent on their own positioning in society, shatters the idea that there is a general, expected way to live illness and that there is *one* experience of illness. For example, if one were to understand living post-stroke as an ongoing sequence of encounters instead of as a set of deficits to overcome, then the experience of living with the effects of stroke would encompass multiple meanings of what it is like to live post-stroke. Encounters with the built environment, various healthcare specialists, the barrage of odd sensations, emotional stress, threatened job and income security and supportive and non-supportive friendships would generate (create) a series of experiences (that form an overall, collective sense of one experience) of illness that are inextricably entwined with one's immediate life. Thus, rather than holding onto a notion of a faulty self-residing in partially functioning body, living post-stroke might well initiate a reconfiguration of self that refuse the idea that there can ever be a universal experience of illness.

Feminists interested in health and illness have taken this relational notion of experience to heart and have written extensively about illness through autobiography (some early works include, e.g., Behar, 1996; Wendell, 1996; Moss, 1999). Several collections exist that have enriched the literature across disciplines by providing autobiographical accounts of living with illness, dealing with the hegemony of medical understandings of the body, and finding political strategies to address the marginalization of and discrimination against persons with ill bodies (e.g., Mattingly and Garro, 2000; DasGupta and Hurst, 2007; Driedger and Owen, 2008).

Most recently feminists have taken up illness through ideas about how discourse (ideas, notions, meaning) and materiality (physical bodies, institutional processes, economic resources) interact and how subjects form through sets of social relations in autobiographies of illness as well as through writing autobiographically (e.g., Alaimo, 2009; Barnett, 2012; Trivelli, 2014). Much of this feminist work draws on new materialism, an emergent body of thought with disparate origins. New materialism can be seen as an amalgamation of strands of thought that seek to explore connectivities and contradictions in the relationship between matter and the language used to talk about it. As a series of reappraisals across social science and the humanities, new materialism is re-entering theory by looking at the materializations of bodies, subjects, and politics (Coole and Frost, 2010). For example, medicine is one language used to talk about ill bodies with regard to atherosclerosis, the protocols in place for diagnosis (variation in blood pressure, poor skin oxygenation, weak pulse, cold feet), the instruments used to assess the physical body (microscopes, ultrasound), and the treatment options (surgery, walking therapy) are as much part of the illness as the bodily sensations of pain, the ability to articulate one's bodily sensations, and access to a medical professional (see Mol, 2002). Embracing an approach that includes both discourse and materiality could lay bare more specifically how it is that individuals experience illness.

Geography

Given that dominance of explorations of experience of illness in disciplines across the social sciences (anthropology, sociology, women's and gender studies) and professional disciplines (education, nursing, social work), one might expect illness narratives to play a large role in framing health geography more generally. But this is not the case. In geography, writing one's illness has been a relatively minor part of health geography with much of the literature existing at the edge of critical approaches in health geography where disability, methodology and innovative analysis overlap (e.g., Longhurst, 2011; Liggins, et al., 2013). As Worth (2008) so clearly shows, however, experience does matter in studies of disability (and illness) and the personal shapes the way research topics are taken up.

Feminist analysis through autobiographical writing is establishing itself in geography through writing one's own illness. In a piece exploring ways to write autobiographically, I attempt in three different ways to link my career path to various points in the destabilization of my body (Moss, 2013). I argue that 'I' is a transient placeholder that momentarily contains that which comes to be a subject. By tracing the shift in topics I researched, the institutional locations I occupied, and my engagement with disability studies as a discipline, I tease out the complexity of what counts as 'I' through various encounters – with friends, colleagues, administrators, and literatures about disability. Depending on the form my writing takes – as a site of inquiry, an expression of relational materiality, or a haecceity – the tracings of 'I' disclose various ways my illness is entwined in the constitution of *a* subject. Bondi (2014), too, tries to figure out how 'I' emerges in a particular place. In an attempt to theorize insecurity, she draws on her own feelings and her experiences with depression and anxiety 'to explore how I sense, narrate and reflect in ways that variously express, perplex, surprise, worry and defy my own changing sense of who I am in the world' (p. 335). In her framework the unconscious is valued and is used to explain that what people say about themselves often hides more significant truths. Through the use of her own emotional distress, she is able to call attention to the paradoxical situatedness of the notion of ontological security (feeling fundamentally, definably safe and intact). Through stories about her own anxiety, she shows that selves are not individually, tightly wrapped packages that populate the world; rather, selves are continuously produced across moments that already include connections with other people through train schedules, academic conferences and a set of personal responsibilities.

Locating her work in the interdisciplinary space of death and trauma studies, Madge (2014) discusses how her own breast cancer permitted her to develop a more embodied notion of compassion and care. Having encountered disembodied, a-emotional and unfeeling accounts of cancer circulating in the health literatures in geography, she uses 'autobiography as a means to attempt to make sense of the perplexing, fleshy, difficulties of the human experience of livingdying from cancer, and the emotional, embodied and political context of that experience' (Madge, 2014, p. 4). She uses the term 'livingdying' to more effectively depict her own experience of breast cancer. She argues that the emotional and physical eclipse of people with ill bodies within academia forecloses insights into the messy reality of people's lives and bodies who are going through diagnosis and treatment of cancer and of the contexts they live and work within, particularly those suspended between life and death. Like Madge, Vanolo (2014) draws attention to the multiple types of space that come to generate the experience of therapy as transformative: emotions, relations, imagination, physicality, power and dreams. He provides a vivid account of what the space of treatment is like during psychotherapy. He draws on notions of therapeutic landscapes and psychoanalysis to conceptualize the room where therapy takes place, as well as the unevenness of how therapy actually works.

Even though experience is accepted as a valid source of knowledge in disability and illness studies in geography, nearly all pieces drawing on autobiography methodologically still provide a justification resembling something to the effect that by using autobiography, the author is not engaging in the task of navel-gazing, but acting as a researcher. Even in critical research approaches – there is still the notion that using one's own experience somehow produces a less rigorous analysis than using other types of qualitative data (Kearns, 2014). Researchers using autobiography as a methodology must depend on critical reflection and precise conceptualizations both to legitimate their work and to produce effective analysis. Critical reflection, entailing the continual querying of assumptions brought to the research, relates to the social relations of knowledge production. Because of the omnipresence of data about one's own experience, including illness, critics sometimes find it difficult to accept an analysis that appears to be based on data that suits the argument. But autobiography as feminist analysis is an intersubjective practice that specifically draws attention to the way knowledge systems circulate and vie for dominance. Analyses act as interventions into this process. Even in feminist circles, feminist guidebooks to writing autobiographically are much less prominent than the outlining methodological problems arising when using one's own experiences as a source of data (for guidebooks, see Ellis, 2004; Muncey, 2010). Again, through critical reflection and precise theoretical conceptualizations, feminists can use autobiography effectively as an analytical tool without getting caught up in the task of justifying the use of their own experience or engaging in the repetitive exercise of reconceptualizing experience.

Becoming, enactment and the material-discursive in diffractive readings

One effective strategy for freeing myself from engaging in the task of justifying my use of autobiography in writing illness has been to use my story to access the processes that generate my 'I' as *a* subject, repeatedly in different contexts with similar yet different formations (after Moss, 2001a, pp. 20–21; 2013, sec. 1, para. 4). I work with concepts that emphasize precariousness, ambiguity and diffidence in addressing issues of subjectivity. Becoming is one of these concepts. I use becoming somewhat like Deleuze and Guattari (1987, pp. 292–293) do; a move away from a dominant subject positioning toward a subjugated one, as in becoming-woman, becoming-insect, and becoming-imperceptible. But I also use becoming as Braidotti (2012, pp. 31, 119, 155–167) does: a description of the process through which a subject comes to be formed in a particular moment, to be invented over and over again, to be porous enough to intervene in as part of an affirmative politics. Becomings are provisional spatial and temporal tracings that indicate rather than define, destabilize rather than fix, and hover around the notion of 'not yet' rather than 'no longer.'

Another concept I find useful is enactment. Enactment dovetails nicely with becoming, for it rests on two interrelated premises: a variety of things come together in a single moment to produce something, and whatever this something turns out to be – someone's life, a subjectivity, an illness – hangs together as an entity. Enact, the verb form, offers a sense of persistent openness to what might come next and belies the fixity of any entity: 'It is possible to say that in practice, objects are *enacted.* This suggests that activities take place – but leave the actors vague. It also suggests that in the act, and only then and there, something *is* – being enacted' (Mol, 2002, pp. 32–33). Through a series of enactments, disclosed through 'techniques that make things visible, audible, tangible, knowable' (Mol, 2002, p. 33), things come to *be* – things come to be *enacted* – that may include a subject in a general sense (the ill body) and a subject in a specific sense (*my* ill body).

In my attempt to write illness through an understanding of subjectivity, I also want to hold onto the complexity of things that compose the body. Feminists, especially those working with new materialisms, recognize the deeply entwined connections between the discursive constructions of ideals and their deviations (via classification through diagnosis) and the matter that gives form to bodies. Matter includes both bodily substance, such as flesh, blood and bones and the systems that connect them, and the material effects of one's positioning in the world (in reference to economics as well as bodily comportment and movement). Some of these feminists use the term 'material-discursive' as a reminder of this deep entwinement and constitutive intra-action between matter and the way we talk about it (see Barad, 2003). Bearing in mind the activity of the material-discursive assists in understanding more precisely the process of enactment, that is, how things – that can be more discursive like a diagnosis or more material like disease – come together to create something that we come to understand as illness and to generate a specific 'I' through a set of practices and activities that are temporary and in flux.

As a set, these concepts offer a way to write illness beyond an idiosyncratic account of being ill, living with illness, or having an ill body. Indeed, using the personal to disclose the workings of wider structures and processes is an integral part of a critical reflective practice (Fook and Gardner, 2007). Yet writing illness through autobiographical writing is also an example of a diffractive reading, which is 'not about what is told or experienced – it is about the ways in which what is experienced is formed' (Jackson and Mazzei, 2012, p. 130). In a diffractive reading, experience is not used as the object of critical reflection; rather, experience emerges as something in need of scrutiny. Writing illness through autobiographical writing makes it possible to position experiences of pain, fatigue and bodily disquiet in light of a set of (material-discursive) practices (expressed as enactments) that disclose some of the processes (becoming) that generate subjects collectively (those with illness) and individually (*my* illness).

Writing my illness – again – in the context of becoming an academic subject

I have written several autobiographical pieces about my own illness: how I practice my scholarship as an academic with chronic illness, how institutional policies and processes shape my emerging subjectivity in the academy, how particular concepts capture the destabilization of my material body and how both the demands in and the organization of the academic workplace facilitate the ongoing generation of my body as ill. Yet I rarely point out how I actually do the analysis (for some discussion see Moss, 2001b; 2012; 2014b). I write about specific academic practices – such as data collection methods, on-campus political organizing, and various types of administrative work – tracing some of the enactments of my illness in light of becoming an academic over the course of my career.

As a bit of background, I developed excessive pain and debilitating fatigue and was diagnosed with ME in 1995 after three years in a hostile workplace, also known as a 'chilly climate', where I was personally harassed and bullied (see The Chilly Collective, 1995; Hannah, et al., 2002). From my own research I knew that the most effective way to treat ME was through managing energy levels. So I did. And I returned at a reduced workload after a year and a half, and returned to full-time in May 1998. I negotiated an accommodation plan with the university administration, the main elements of which included one course reduction per year, restricted on-campus engagement, and work stations at my campus and home offices. Through 'pacing' as an energy management strategy I was able to maintain full-time employment. As my energy levels stabilized, I began to push myself to do more academic work. I took on more book-based research and administrative tasks, for this was the kind of work I could do without taxing my energy levels. I was wrong. I had a major relapse in 2007. My brain stopped working. I couldn't read, I couldn't hold a conversation for more than minute or two, and I couldn't write. Many with ME live with this kind of extreme cognitive impairment; I just had not been one of them. It took about a year for me to regain my thinking skills. That incident made me more determined to stick to pacing.

Writing illness through my autobiographical writing, using critical reflection while generating a diffractive reading, requires more than just highlighting a few bodily breakdowns both as an effect of academic practice and as an element in the way in which my academic practice takes shape. After rereading my autobiographical writings as a set, I see that I equivocate on the many things that produce me as an academic subject in various settings: at home, in university processes, and at conferences. In the rest of this section, I want to write my illness by using some fairly mundane examples in each setting to show how becoming, enactment and the material-discursive can accentuate the precarious, ambiguous, and diffident aspects of the generation of 'I' as an academic subject.

Part of being ill with ME means getting incapacitating headaches. Mine start at the top of my head and slowly envelope my temples, base of my skull, and the tops of my shoulders. Unlike a migraine, where I curl up in a dark room and lie motionless, with an ME headache I want to shake my head to get the pain out. Yet any movement creates a shooting pain, just as stillness builds up pressure. I do not remember much of what I do when I have one of these headaches. I do know that when it goes away – it stops all of a sudden and leaves no echo of a presence – I feel both vanquished and revitalized: After utter prostration, I begin feeling energy seeping back into my body one cell at a time.

I live in fear of an ME headache. My partner gets a daily report, augmented with various bulletins and flashes throughout the day, when I feel that something is a little off. A typical report would be something along the lines of:

> I have a headache, but not the sinus-y headache I had the day before yesterday, it is more like the one I have on the top of my head. I had one of those when we were playing pool with Herbert. Do you remember? I went home and couldn't make it up the steps. You walked up with me. It took over a half an hour to go up three flights. Remember? Didn't I lay in bed until the next afternoon? And then it lifted. As if it had never been there at all? It's not as bad as that one – obviously – because I'm sitting here in the kitchen. But it could be.

Then I would make a sound, barely audible, that was an expression of a combination of bewilderment, fret and disfavor.

Most days if I were feeling like this, I would make my way to my home office. It is a room solely for work – nothing else. On a day where there is the possibility of a headache, I try to write because of the intense focus, which holds the pain at bay. A bulletin later in the day, maybe when I break for lunch, would be something like: 'I've still got it. It's kind of fading. But I feel the echoes when I look down at the keyboard. I don't think it is bad. I hope it isn't.' I would be scrunching up my nose just to check if I had been wrong, and it was a sinus-y headache after all. Then I would report on that. And then later, after a Skype meeting or a stretch of marking assignments, I might report something like, 'I'm feeling better. It wasn't an ME headache.' I would be relieved, and in that relief, I would push back the concern of its return tomorrow – for the headaches happen in clumps, and it could actually be a sign that one is coming – and finish my work for the day.

Talking my illness like this helps me frame both my working space at home and me as an academic subject. Entering my office and closing the door sets me up to engage with academic work. After having talked about the potential destabilization of my body, I feel somehow formed enough for the day whereby I have some parameters, albeit precariously laid out, within which I can do the work that I am supposed to do. Through this talk I momentarily enact myself as an academic subject while negotiating the material boundaries of becoming ill. I can see that it is incredibly important for me to be precise in the type of pain I feel, to convey my sensations to someone who has been part of my life, who has been part of talking (and now writing) my illness with me. This talk produces innumerable categories that I reference with ease, reassuring myself that I know – I know my pain, my body, my illness – alongside a faint awareness that I really do not know. This talk both pulls me along and propels me in my own becoming ill, well, healthy and 'I'. This talk puts words to my pain, my feelings, my living. Through this material-discursive connection I come to be aware of the volatility of my becoming an academic subject who is both ill and well at the same time, much like Madge (2014) is aware of her positioning as livingdying.

Given this rather uneven and precarious engagement in academic work, the practice of annual evaluations, where the documentation of excellence matters, draws attention to static notions of what academic subjects need to be. Excellence rests on the ability to demonstrate productivity through high numbers in publications and research dollars. It is not just any publication or research grant; within each category there is a hierarchy determined by a set of criteria that values peer evaluation, national and global connections and external resourcing. Self-promotion is the strategy used to write excellence, to talk an academic subject. The neoliberal academic subject (see Davies, 2006; Davies, et al., 2005) exalted in these processes is able-bodied and healthy (following Berg, Gahman and Nunn, 2014). Even though policies in place that direct administrators to take into account chronic illness, disabling illness, and disability, the effects of the assessment reinforce the model of the academic subject as one that is robust, healthy, forceful, and vigorous in the customary academic practices of publishing and securing grants.

My own ill body contributing to the generation of 'I' as an academic subject disrupts this solidity, yet my submission reinforces the neoliberal sentiment of the practice of requiring an annual report. My documentation notes that my accomplishments for the review period are to take into account the fluctuations in health. I produce an image of my body as one that is broken, a little needy and requires special attention. This rights-based orientation (Rioux and Valentine, 2006) of the evaluation process forces me to enact a subject positioning that claims as legitimate a series of concessions so that I may be considered equal to my colleagues. In writing illness into my professional life through annual evaluations, I am cut off from providing an analysis of what effect this particular orientation has to my academic practice and to me as an academic subject. I am unable to explain the extant material-discursive relationships that comprise my working environments and shape what academic practices I engage in. I am refused the opportunity to talk about becoming as a process through which I am at the same time practicing illness *and* health, becoming a wandering subject *and* ascribing to a hegemonic one, and enacting academic practices in a generative *and* a static way. In short,

I am refused to talk my illness in relation to being an academic subject in the way I experience illness and my academic subjectivity.

Juxtaposing talking my illness as a way to set some parameters so that I can pace myself against the strictures of an evaluation practice in the university indicates two oppositional practices that enact me as a specific academic subject in quite diverse ways. The first is a generative, substantially unstructured practice through which I temporarily assemble activities that attempt to prevent deterioration whereas the second is a restrictive, highly structured activity through which a specific subject positioning is imposed upon a neatly arranged set of achievements. Such a distinction is not always as clear as I make it out to be. For example, at conferences, I bounce back and forth between the two to a point that leaves my head spinning. As part of my pacing strategy, I plan which conferences I go to. In those plans, I try to make room for being ill as I engage in academic work outside my usual spaces: I stay by myself, I schedule lots of time between activities, and I minimize interaction in large groups. Even so, I'm still drawn to talking with colleagues, embracing the serendipitous directions conversations take us and finding out what other people are thinking – the stuff that makes conferences worth going to. This is exactly the stuff that drains my energy, facilitates disruptions to systems in my body and pokes holes in my parameters that I so carefully set up to prevent larger, more encompassing crashes.

The notions and ideas that I have about being an academic at a conference are supported by and constitutive of the practices I engage in. It is a place for networking, forging collaborations and being energized by the exciting work that academics are doing. But to keep myself sensibly intact, to ward off a crash for which I would take months to recover, means doing the bare minimum – presenting my work and then leaving the conference. This past year, although I tried to gauge the impact of my actions, I misjudged my thresholds, and I tended to outpace myself. I attended four conferences across four months, and it took about six and a half months to get back to the (compromised) energy levels I had before the first one. Although I worked diligently on a daily basis to talk my parameters aloud so that I would know when I crossed a threshold, at a larger scale, I ended up breaking apart the boundaries that held together my becoming. I ended up enacting an academic subject that was partly generative and partly restrictive, one that continually negotiated the tension between the capacity of an ill body to engage in academic activities and what counts as excellence in an academic setting.

Closing thoughts about writing illness

Writing illness as a critical approach in health geography involves some critical engagement with people's experiences of illness. Information about living with illness can be gathered through a variety of methods including interviews, diaries, and surveys and cover topics as diverse as the experience of the breakdown of the body, the ease of access to healthcare services or the decision to refuse standard treatment protocols. Writing illness through feminist autobiographical analysis is a critical methodology that generates diffractive readings that underscore how the experience of illness itself forms. By focusing on *how* experiences form and the context within which particular experiences emerge, feminist autobiographical analysis goes beyond the taken for granted discourses that inform much of the literature on the experience of illness and demonstrate more soundly what gives rise to both these common narratives and alternative ones.

As an illustration of how writing illness through feminist autobiographical writing can work, I engaged with processes that produce an academic subject, one that appears healthy but always in tune with bodily sensations that could indicate episodes of more intense illness. Living with seemingly aleatory fluctuations of bodily processes, the turmoil of becoming both ill and an academic subject positions me as someone with ME as living on a threshold, within the limit of illness. Talking the boundaries of the body is a practice that seeks to tease out some of the significant elements that comprise the material-discursive context of any given moment – whether it be at the scale of now, a trend over the last few days or the accumulation of elements over a period of months. This talk is part of pacing that helps me keep tabs on the limits of my body so that I won't crash, that is, go beyond an energy threshold that would initiate changes in multiple bodily systems that have varying recovery times. Exploring the planning of how to engage with the context that gives rise to how I experience illness in practices that produce me as an academic subject (daily writing, annual reviews and attending conferences) is one way to disclose aspects of living within a threshold. Without the constant self-monitoring, without the talk, without the engagement with particular people and spaces, these thresholds appear as uniform boundaries, invoked by health professionals charged with treating ill bodies, as in, 'Don't get tired', and 'Don't over-do it.' But it is this living within the threshold that forms my experience of living with an illness and being an academic.

Writing illness through feminist autobiographical analysis addresses a specific set of research questions, ones oriented toward experiences of illness. Experience here is conceptualized as an active force – not only in constituting a subject and the illness, but also in shaping the context within which practices take place, a conception much different than more traditional humanist understandings of experience as something fundamental to the human condition. Experience is still a valued form of knowledge that informs how it is that specific subjects emerge. Using this notion of experience facilitates a more fine-grained analysis of the temporarily fixed links between illness and subjectivity. Such an analysis can help show what sorts of subjects emerge through circumscribed sets of practices. It is also an effective way to make sense of the complexity of your own experience of illness!

References

Alaimo, S., 2009. MCS matters: material agency in the science and practices of environmental illness. *TOPIA: Canadian Journal of Cultural Studies*, 21, pp. 9–28.

Angus, J. 2008. Contesting coronary candidacy: reframing risk modification in coronary heart disease. In Moss, P., and Teghtsoonian, K. (eds.) *Contesting illness: processes and practices*. Toronto: University of Toronto Press. pp. 90–106.

Barad, K., 2003. Posthumanist performativity: toward an understanding of how matter comes to matter. *Signs*, 28(3), pp. 801–831.

Barnett, T., 2012. Remediating the infected body: writing the viral self in Melinda Rackham's *Carrier*. *Biography: An Interdisciplinary Quarterly*, 35, pp. 45–64.

Behar, R., 1996. The girl in a cast. In *The vulnerable observer: anthropology that breaks your heart*. Boston: Beacon Press. Ch. 4, pp. 104–135.

Berg, L., Gahman, L., and Nunn, N., 2014. Neoliberalism, masculinities and academic knowledge production: towards a theory of 'academic masculinities'. In Gorman-Murray, A., and Hopkins, P. (eds.) *Masculinities and place*. Burlington, VT: Ashgate, 58–75.

Bondi, L., 2014. Feeling insecure: a personal account in a psychoanalytic voice. *Social and Cultural Geography*, 15, pp. 332–350. doi:10.1080/14649365.2013.864783.

Braidotti, R., 2012. *Nomadic theory: the portable Rose Braidotti*. New York: Columbia University Press.

Brodzki, B. and Schenck, C., eds., 1988. *Life/lines: theorizing women's autobiography*. Ithaca, NY: Cornell University Press.

Bury, M., 1982. Chronic illness as biographical disruption. *Sociology of Health & Illness*, 4, pp. 167–182. doi:10.1111/1467-9566.ep11339939.

Castleden, H., Crooks, V.A., Schuurman, N. and Hanlon, N., 2010. 'It's not necessarily the distance on the map . . . ': using place as an analytic tool to elucidate geographic issues central to rural palliative care. *Health and Place*, 16, pp. 284–290. doi:10.1016/j.healthplace.2009.10.011.

The Chilly Collective, eds., 1995. *Breaking anonymity: the chilly climate for women faculty*. Waterloo, ON: Wilfrid Laurier University Press.

Coole, D., and Frost, S., 2010. Introducing the new materialisms. In Coole, D., and Frost, S. (eds.) *New materialisms: ontology, agency, and politics*. Durham, NC: Duke University Press. pp. 1–43.

Coslett, T., Lury, C., and Summerfield, P. (eds.), 2000. *Feminism and autobiography: texts, theories, methods*. New York: Routledge.

Crooks, V. A., Agarwal, G., and Harrison, A., 2012. Chronically ill Canadians' experiences of being unattached to a family doctor: a qualitative study of marginalized patients in British Columbia. *BMC Family Practice*, 13(69). doi:10.1186/1471-2296-13-69.

Crooks, V. A., Hynie, M., Killian, K., Giesbrecht, M. and Castleden, H., 2011. Female newcomers' adjustment to life in Toronto, Canada: sources of mental stress and their implications for delivering primary mental health care. *GeoJournal*, 76, pp. 139–149. doi:10.1007/s10708-009-9287-4.

Curtis, S., 2010. *Space, place and mental health*. Burlington, VT: Ashgate.

DasGupta, S., and Hurst, M., 2007. *Stories of illness and healing: women write their bodies*. Kent, OH: Kent University Press.

Davidson, J., and Henderson, V. L., 2010. 'Coming out' on the spectrum: autism, identity and disclosure. *Social & Cultural Geography*, 11, pp. 155–170. doi:10.1080/14649360903525240.

Davies, B., 2006. Women and transgression in the halls of academe. *Studies in Higher Education*, 31(4), pp. 497–509. doi:10.1080/03075070600800699.

Davies, B., Browne, J., Gannon, S., Honan, E., and Somerville, M., 2005. Embodied women at work in neoliberal times and places. *Gender, Work and Organization*, 12, pp. 343–362. doi:10.1111/j.1468-0432.2005.00277.x.

Deleuze, G., and Guattari, F., 1987. *A thousand plateaus: capitalism and schizophrenia*. Minneapolis: University of Minnesota Press.

Donovan, C., 2014. Representations of health, embodiment, and experience in graphic memoir. *Configurations*, 22, pp. 237–253. doi:10.1353/con.2014.0013.

Driedger, D., and Owen, M., eds., 2008. *Dissonant disabilities: women with chronic illness explore their lives*. Toronto: Women's Press.

Ellis, C., 2004. *The autoethnographic I: a methodological novel about autoethnography*. Lanham, MA: Altmira.

Fook, J., and Gardner, F., 2007. *Practising critical reflection: a resource handbook*. Berkshire, UK: McGraw-Hill International.

Foucault, M., 1990. *History of sexuality*, vol. 1, *An introduction*. New York: Vintage Books.

Frank, A., 1991. *At the will of the body: reflections on illness*. New York: Houghton Mifflin.

Gilmore, L., 1994. *Autobiographics: a feminist theory of women's self-representation*. Ithaca, NY: Cornell University Press.

Hannah, E., Paul, L., and Vethamany-Globus, S., 2002. *Women in the Canadian academic tundra: challenging the chill*. Montréal-Kingston: McGill-Queen's University Press.

Jackson, A. Y., and Mazzei, L. A., 2012. *Thinking with theory in qualitative research: viewing data across multiple perspectives*. London: Routledge.

Kearns, R. A., 2014. The health in 'life's infinite doings': A response to Andrews et al. *Social Science & Medicine, 115*, pp. 147–149. doi:10.1111/j.0033-0124.1993.00139.x.

Liaschenko, J., Peden-McAlpine, C., and Andrews, G. J., 2011. Institutional geographies in dying: Nurses' actions and observations on dying spaces inside and outside intensive care units. *Health and Place*, 17, pp. 814–821. doi:10.1016/j.healthplace.2011.03.004.

Liggins, J., Kearns, R. A., and Adams, P. J., 2013. Using autoethnography to reclaim the 'place of healing' in mental health care. *Social Science and Medicine*, 91, pp. 105–109. doi:10.1016/j.socscimed.2012.06.013.

Longhurst, R., 2012. Becoming smaller: autobiographical spaces of weight loss. *Antipode*, 44, pp. 871–888. doi:10.1111/j.1467-8330.2011.00895.x.

Madge, C., 2014. Living through, living with and living on from breast cancer in the UK: creative cathartic methodologies, cancerous spaces and a politics of compassion. *Social and Cultural Geography* (pre-publication, December 16, 2014), pp. 1–26. doi:10.1080/14649365.2014.990498.

Mattingly, C., and Garro, L. C. (eds.) 2000. *Narrative and the cultural construction of illness and healing*. Berkeley: University of California Press.

Miller, N. K., 1991. *Getting personal: feminist occasions and other autobiographical acts*. New York: Routledge.

Mol, A., 2002. *The body multiple: ontology in medical practice*. Durham, NC: Duke University Press.

Moss, P., 1997. Negotiating spaces in home environments: older women living with arthritis. *Social Science and Medicine*, 45(1), pp. 23–33. doi:10.1016/S0277-9536(96)00305-X.

Moss, P., 1999. Autobiographical notes on chronic illness. In Butler, R., and Parr, H. (eds.) *Mind and body spaces: geographies of disability, illness and impairment*. London: Routledge. pp. 155–166.

Moss, P., 2000. Not quite abled and not quite disabled: experiences of being 'in between' ME and the academy. *Disability Studies Quarterly*, 20(3), pp. 287–293.

Moss, P., 2001a. Writing one's life. In Moss, P. (ed.), *Placing autobiography in geography*. Syracuse: Syracuse University Press. pp. 1–21.

Moss, P., 2001b. Engaging autobiography. In Moss, P. (ed.), *Placing autobiography in geography*. Syracuse: Syracuse University Press. pp. 188–198.

Moss, P., 2003. Re-reading ill bodies as healthy bodies: positive ontology in practice. *Chimera*, 18, pp. 111–115.
Moss, P., 2008. Edging embodiment and embodying categories: reading bodies marked with Myalgic encephalomyelitis as a contested illness. In Moss, P., and Teghtsoonian, K. (eds.) *Contesting illness: processes and practices*. Toronto: University of Toronto Press. pp. 158–180.
Moss, P., 2012. Taking stock in the interim: the stuck, the tired, and the exhausted. *Antipode*. Available at: http://antipodefoundation.org/2012/10/15/symposium-on-the-participatory-geographies-research-groups-communifesto-for-fuller-geographies-towards-mutual-security/#more-1970. (Accessed January 21, 2015.)
Moss, P., 2013. Becoming-undisciplined through my foray into Disability Studies. *Disability Studies Quarterly*, 33(2), n.p. Available at: http://dsq-sds.org/article/view/3712/3232. (Accessed 21 January 2015.)
Moss, P., 2014a. Shifting from nervous to normal through the love machine: battle exhaustion, military psychiatrists and emotionally traumatized soldiers in World War II. *Emotion, Space and Society*, 10, pp. 63–70. doi:10.1016/j.emospa.2013.04.001.
Moss, P., 2014b. Some rhizomatic recollections of a feminist geographer: working toward an affirmative politics. *Gender, Place and Culture,* 21(7), pp. 803–812. doi:10.1080/0966369X.2014.939159.
Moss, P., and Dyck, I., 2002. *Women, body, illness: space and identity in the everyday lives of women with chronic illness*. Lanham, MA: Rowman & Littlefield.
Muncey, T., 2010. *Creating autoethnographies*. Thousand Oaks, CA: Sage.
Reiser, S. J., 2009. *Technological medicine: the changing world of doctors and patients*. New York: Cambridge University Press.
Rioux, M., and Valentine, F., 2006. Does theory matter? Exploring the nexus between disability, human rights, and public policy. In Pothier, D., and Devlin, R. (eds.), *Critical disability theory: philosophy, politics, policy and law*. Vancouver: University of British Columbia Press, pp. 47–69.
Scott, J. W., 1991. The evidence of experience. *Critical Inquiry*, 17, pp. 773–797.
Smith, S., and Watson, J. (eds.), 1998. *Women, autobiography, theory: a reader*. Madison: University of Wisconsin Press.
Sultana, F., 2012. Producing Contaminated citizens: toward a nature-society geography of health and well-being. *Annals of the Association of American Geographers*, 102, pp. 1165–1172. doi:10.1080/00045608.2012.671127.
Thien, D., and Del Casino, V. J., 2012. (Un)healthy men, masculinities, and the geographies of health. *Annals of the Association of American Geographers*, 102, pp. 1146–1156. doi: 10.1080/00045608.2012.687350.
Trivelli, E., 2014. Depression, performativity and the conflicted body: an auto-ethnography of self-medication. *Subjectivity*, 7, pp. 151–170. doi:10.1057/sub.2014.4.
Vanolo, A., 2014. Locating the couch: an autobiographical analysis of the multiple spatialities of psychoanalytic therapy. *Social and Cultural Geography*, 15, pp. 368–384. doi: 10.1080/14649365.2014.882973.
Wendell, S., 1996. *The rejected body: feminist philosophical reflections on disability*. New York: Routledge.
Worth, N., 2008. The significance of the personal within disability geography. *Area*, 40, pp. 306–314. doi:10.1111/j.1475-4762.2008.00835.x.

6 Community capacity building through qualitative methodologies

Sarah A. Lovell and Mark W. Rosenberg

Reflections on praxis[1]

My (Sarah A. Lovell) decision to adopt community-engaged research was driven by an interest in using research to achieve goals of social justice. While conducting interviews and working with the results of my master's thesis, I had become acutely aware of how one-directional the relationship between participants and researchers can be in traditional research. The feminist methodologies I had adopted suggested that the interview process should be an empowering experience for the participant as the researcher in valuing and validating the participant's perspectives. However, like many geographers who have adopted activist and participatory approaches, I found this to be a fairly shallow level of empowerment and commitment to the participant.

Methodologically, participatory research is loose and evolving which allowed me the freedom to adopt feminist strategies of positionality and reflexivity that seek to make the research process more transparent and enhance the quality of research being produced. In practice, this involved explicitly reflecting on the viewpoints I brought into the research and my own social distance and interconnections with the research problem and 'participants'. These practices of positionality and reflexivity were not overly personal but did acknowledge how my beliefs and motivations shaped the research and, in this way, were ultimately political in questioning my role in society and academia. I discuss these experiences in this 'reflection on praxis', whereas the remainder of the chapter is co-authored.

While academics have discussed the challenges of undertaking participatory research and its institutional incompatibilities, few have examined how we, as researchers, experience the loss of control that comes with genuine community collaboration. Community-based participatory research approaches advocate community control over the research problem and its formulation. I entered into the research process concerned that the community would choose a research topic that was beyond my area of expertise. It was with some relief that the community decided to address the shortcomings of the city's accessible transit service, a problem that encapsulated the focus of health geography to me. The subsequent research experience illustrates that engaging communities in research can help marginalized groups to feel empowered and build social support. Working with

1 This section is written by Sarah A. Lovell.

a group of disabled individuals to address the accessible transit service provided incredible insight into the everyday challenges users faced and how these could be compounded by chronic conditions and frequent hospital visits. Bus users, for example, found it difficult to predict when their appointments or events would end and the possibility of missing their bus or a long wait caused stress for many, as one participant explained: 'I find you just guess what time you think you might be done at and pray to God it's done by the time it shows. . . . And I know exactly what you're talking about– it's an absolute nightmare.' Underpinning their experiences of disability were common themes of finite resources and unmet need. While I felt these problems were incredibly important, I also realized that the topic may have limited appeal to academic journals. Throughout the research process I remained unsettled as to how my role as an academic and producer of knowledge would develop if I permanently relinquished control over my own research agenda to communities. This was particularly evident when the research stakeholders came to me and said they wanted to address the accessibility of the local shops for our next project.

While no doubt subject to debate, I see the primary role of academic researchers to be the furthering of societal knowledge. Addressing the needs of communities is an infinitely worthwhile task; however, I could not help but think that this is a role already well occupied by government, non-profits and volunteers, so should academics be replicating this work? Indeed, throughout the research there were moments and days where we had the uncomfortable feeling that a professional in the field could navigate the process of community engagement or meet the needs of disabled stakeholders better than us. Some years later, I co-presented a seminar on participatory research with a colleague who had worked for years as a health promoter. The community engagement practices that I had self-consciously employed and fretted over barely received a mention in his presentation as they were second nature to the way he worked every day.

Participatory research sits uncomfortably next to dominant research paradigms in the social sciences as its collaborative nature produces results that cannot be easily assigned to any one individual. The researcher is unable to define clearly their input into the process or claim authorship over publications, thereby negating mainstream expectations of a successful academic. This is heightened in circumstances where the requirements of a qualification, such as a PhD, are being met, as most state that research must be owned and attributed to one individual who is being evaluated on his or her ability – not a community's – to produce research. Kesby (2000, p. 432) raises the question of how we can conduct praxis-oriented work, 'while at the same time meeting the criteria which [our] embeddedness in an audit-oriented academy dictates.'

In the years since completing the research, qualitative health geography has become increasingly theoretical in its analysis of data, further widening the chasm between the public life of a community-engaged researcher and the academic obligations of a scholar. I wondered how we were going to employ French theorists to interpret our research on access to transportation for disabled persons in order to get published in *Area* and what the stakeholders would think if we proposed it to them! Indeed, the increasingly audit-oriented academy necessitates we must engage in such theoretical debates in order to meet publication outcomes as we are, as Kearns and Moon (2002) argue, as much shaped by neoliberalism as the people we study. In contrast to the work discussed in this chapter, I have chosen to address this problem by embracing more of a consultative approach to community engagement in research and increasingly believe that developing my own, narrower, area of expertise will enhance the contribution that we can make as academics. I see the importance of community involvement in research, but we also recognize the reward that comes with determining one's own research agenda and interests.

Introduction

Harnessing the skills, networks and resources of a community to achieve positive change has long been a focus of academic research (Freire, 1970; Labonte, 1997). In recent decades, discussion has turned to examine whether the research process itself may be a tool for engaging and empowering communities. The rise of participatory methodologies has brought into question some of the fundamental assumptions underlying research in geography and demanded that we take a critical look at the distribution of power upon which academic research is based. Health geographers have responded to this shift by adopting participatory methods such as body mapping and photovoice (Bijoux and Myers, 2006; Kesby, 2004; Castleden, et al., 2008) and a new ethic of partnership with communities (Kitchin, 2001; Wilton, 2004; Tobias, et al., 2013). This chapter examines whether building community capacity, that is, the skills and abilities of a group to address their own health problems, is an achievable outcome of research. In the process, we scrutinize the role of participatory methods in health geography and address key questions facing academic researchers, principally: what is the best strategy for engaging a community in research? How do we share control over research and how can non-academics benefit from being involved in the research process?

Underpinning the discussion are our own experiences engaging communities in research through participatory approaches. Motivated by an interest in achieving social justice and enhancing the experience of research participants, we adopted a model of community-based participatory research (CBPR) to work with users of an accessible transit services for people with disabilities. Through CBPR, community members seek tangible research outcomes while the momentum of social engagement is drawn on to transform society (Park, 1993). When successful, the approach may empower individuals and communities and raise social awareness, leading to tangible community change. Grounded in the work of Freire (1970), CBPR seeks to work with 'oppressed' people to transform their realities and uses research to do this (Maguire, 1987).

In our research, we sought to build community capacity through an intensive model of community engagement (consistent with Cornwall's co-learning level of participation, see Table 6.1). Over an intense six month period, I (Sarah Lovell) worked with the stakeholders as they reflected on their experiences of disability and the challenges of achieving social change. Drawing on the results of interviews and participant observation, we critically examined the experiences of both the academic researcher and community stakeholders as they undertook participatory research for the first time. Central to any form of community engagement is the identification of a goal or issue that individuals can mobilize around. In this project, the disabled stakeholders sought to improve the quality of the local government funded accessible transit service that operates in parallel to mainstream public transit. Despite embarking on the project seeking social change, stakeholders shared a level of cynicism over how effective their efforts would be while so many negative public attitudes toward disability persist. Stakeholders were aware that while we could fight for organizational or institutional change, until the attitudes of the broader public shifted an accessible city would remain elusive. Passion for the issue and the potential to shape public attitudes united and motivated the stakeholders and drove the research process, as the quotes below illustrate:

> June: I think you'd have to do a lot of talking because some people just don't understand it . . . the stuff that's got to be done for handicapped people. . . . Well, I don't think there's going to be any solution unless people get more aware of what the problem is and I don't think that's going to happen right away.
>
> Alice: Well I've found, since I had my leg off, that some places that you go people just . . . they can't be bothered with you. They won't get out of your way, they won't do anything to make things easier for you and [then] other people can't do enough for you.

In this chapter we reflect on the successes achieved through the participatory process and the common challenges that arise for both academics and community members undertaking community-engaged research. We begin the chapter by reviewing how health geographers have engaged with communities in their research and the recent explosion of participatory methods. We subsequently interrogate the spectrum of participatory approaches and the principles that dictate the importance of community engagement to the research process. As a whole, this chapter seeks to examine how health geographers have sought to achieve community engagement in their research and the implications for participants.

Questions of epistemology and community engagement

In this section we begin situating participatory approaches by briefly reviewing the history and methodological developments that have shaped the subdiscipline. Goals of social justice have underpinned research in medical geography since its inception, principally through the identification of groups deemed vulnerable due to their limited access to health services or exposure to environmental health threats (Rosenberg, 2013). This ideological commitment to social justice saw medical geographers regularly engage with communities using research problems to address their problems. Historically, this engagement went unacknowledged in research publications as the subdiscipline's epistemological commitments favored the researcher as an impartial observer.

With their use of survey and secondary data, medical geography was grounded in positivist epistemologies that shaped both the research focus on disease ecology and health services research. Epistemology is broadly defined as a theory of knowledge; epistemologies are a construction of what we accept knowledge to be and what valid sources of knowledge are and should be (Potter, 1999). Positivism is a perspective through which knowledge is understood to be objective, universal and countable and, as such, is deeply entwined with principles of empiricism (Darlaston-Jones, 2007). As a theory, empiricism supports that knowledge that is derived from what can be observed through the senses and draws from 'scientific' methods such as validated survey instruments and experiments to allow for verification through repeat observations of the same phenomenon. This approach lies in contrast with qualitative methodologies where the researcher is a subjective influence on study outcomes. The principles of objectivity underpinning empiricism led collaborations between academics and outside groups to go largely unnoticed in research outputs over the 1980s and much of the 1990s. As health geographers began to distance themselves from the 'parent' discipline of medical geography, the positivist underpinnings of previous research became the subject of criticism from a range of sources. In 1991 Bennett claimed that '[m]edical geographers have seldom recognized the deeper philosophical significance of their technical and methodological concerns' (p. 339). Brown and Duncan (2002, p. 361) further argued that medical geography was seen as 'reductionist, determinist, essentialist and, above all, profoundly a(nti)-social.' This discontent signaled the opportunity to pursue a new geography of health outside of the shadow of epidemiology and the strong influence of medicine (Kearns and Moon, 2002); in turn, this has led researchers to engage with communities in more meaningful ways and question the role of the researcher as impartial observer. The rise of qualitative methodologies, in contrast, positions the researcher as an instrument who can shape the outcomes of a study through the questions she asks and even her body language.

As researchers, our personal experiences and connections often allow us to occupy an insider status that can expose us to new research problems or open up access to a research population that may otherwise have been protected by gatekeepers. The importance of these connections are acknowledged more explicitly today; for example, Kearns and colleagues (2003), discuss their connections as primary school board member and parents that facilitated research into the 'walking school bus' that accompanies children to and from their local primary school. Kearns' voluntary work with the school assisted them in accessing the research community, but it also positioned him as an insider, enabling the research team to engender shared experiences with participants, as they explain: 'A pre-existing knowledge of the school command[ed] an ability to talk with, rather than about, those involved' (Kearns, et al., 2003, p.288). Insiders share a common identity or experience with research participants that facilitates social acceptance and can change the dynamics of research interactions (Hay, 2000). Such methodological observations have been critical to the establishment of more self-aware and engaged health geographies.

The rise of critical perspectives seeking to address the inequalities experienced by marginalized groups in society, such as women and ethnic minorities, led researchers to further question the power relations within research and the reductionist approaches of positivist methods (Maguire, 1987). Movements such as feminism and poststructuralism have strongly influenced the kinds of questions health geographers have been asking and the methods used to answer them. Feminist geographers, for example, have highlighted the tensions that can arise carrying out qualitative research and the need for flexible, responsive, and sensitive researchers (Gibson-Graham, 1994; Davidson, 2001). Feminist researchers seek to undertake research that values participants and questions existing power structures, using qualitative methods as a means of attending to these concerns. Interviews, for example, are widely acknowledged as a means of reducing the power imbalance between the researcher and participants. 'It is suggested that this type of research allows the development of a less exploitative and more egalitarian relationship between a researcher and her participants than is possible in other methodological frameworks' (McDowell, 1992, p. 406). Today, feminist researchers are engaging with emancipatory approaches and research collaborations that empower women across academia and the community (Gilbert and Masucci, 2008).

Geographers are increasingly self-identifying as activists and discussing in the academic literature what this extension of their professional identity means for them as scholars and as members of communities outside of the academy (Askins, 2009; Taylor, 2007). There is no doubt that the increasingly audit-oriented nature of the academy has played a part in the extension of reflexive interrogation and writing up of *all* aspects of our academic roles. Yet transformative research remains largely at the peripheries of academia. The dramatic growth in activist-scholarship occupies what Routledge (1996) terms a 'third space' for 'critical engagement' which connects the academy with the activist world where geographers contribute to causes and community organizations while constantly reflecting on how their own positions impacts – and is impacted by – these very different worlds. Routledge describes the tensions of such an occupation:

> In my own experiences it is not clear to me where one 'role' or position begins and where the other ends. This blurring holds out the possibility that 'insider' and outsider' voices may coalesce into a new perspective, one which is not just counter-hegemonic or simply oppositional (thereby remaining within the discursive frameworks and structures of the dominant), but which opens a new arena for negotiation, meaning, representation. (Routledge, 1996, p. 414).

Criticisms of the normative performance of activism have made some geographers, such as Askins (2009), reluctant to place themselves on the academic-activist landscape. Embarking on a process of transformative research was daunting for us as it relies on the researcher to be intensely self-reflective of their own influence over the process of community engagement and the research. Despite the 'interwoven' nature of activism and academia, due to the social relevance of our work, health geographers have been slow to engage with personal narratives of activism (Andrews, et al., 2012; Taylor, 2014). Dorling, one of a few self-described activists, is noted for crossing the public-academic divide through both his writing and protest work on inequalities and health and, as such, occupies a space dominated by medically trained researchers (Dorling, 2014). Navigating landscapes of research and activism requires flexibility and responsiveness, as the researcher balances community engagement, group facilitation and research skills.

While collaboration can take many forms, increasingly geographers have come to explore the transformative potential of engaging with communities through research. The rapid rise of participatory methods in social geography has led to texts devoted to participatory geographies (Kindon, et al., 2005), and progress reports on the subject in *Progress in Human Geography* (Pain, 2004). Articles devoted to participatory methods have become commonplace in select journals such as *Area* (e.g., Mistry, et al., 2014). Within health geographies, the use of participatory approaches to enhance traditional qualitative methods has increased exponentially with the use of photo diaries and photo voice (Myers, 2010; Thomas, 2007; Castleden, et al., 2008; Ortega-Alcazar and Dyck, 2011), go-along interviews (Oliver, et al., 2011), participatory video (Richmond and Fortier, 2013), and participatory mapping (Kesby, 2004). The smaller body of participatory literature that has sought to use research to build community capacity (e.g., see Kitchin, 2001; McFarlane and Hansen, 2008; Wilton, 2004) and contributed an understanding of the challenging and unpredictable nature of community-engaged research. The following sections draw on the lessons of these researchers and the wider body of participatory literature to critically discuss the goals and scope of such research in the geographies of health.

Approaches to building community capacity through community-engaged research

The use of collaborative research was first documented in developing countries with the purpose of overcoming the cynicism of local communities toward western development initiatives (Green and Mercer, 2001). Despite being grounded in moral ideals of empowering marginalized peoples, the approach was more a reaction by academics to growing resistance amongst overresearched communities than a genuine attempt to improve the ethics of research. In developed countries, the rise of participatory research followed a similar trajectory as Indigenous peoples began to resist the implementation of research where they received only minimal direct benefits from participating, often not even receiving study results (Green and Mercer, 2001). Participatory research has rapidly gained in popularity with a new ethic of engagement whereby researchers are mindful to systematically consult with the public in the design of research

(McFarlane and Hansen, 2008). Expectations of collaboration in participatory research have changed over time and taken on different forms across research projects, places and countries. For example, in countries such as Canada, participatory approaches are increasingly adopted in work with Indigenous populations as a means of redressing power imbalances and undertaking culturally sensitive research (Tobias, et al., 2013; Castleden, et al., 2008). In contrast, New Zealand's research culture has institutionalized processes of Indigenous consultation and collaboration and, increasingly, is extending beyond participation toward an ethic of research 'by Māori, and for Māori', that is undertaken by Indigenous researchers embracing a Māori world view that incorporates Māori control over research and commitment to Māori protocols. Such 'Kaupapa Māori' approaches seek to build the Māori research workforce and ensure the outcomes of research benefit Māori (Walker, et al., 2006).

Underpinning many transformative methodologies is an assumption that the more collaborative a project is the greater the opportunity for community learning. Table 6.1 depicts the range of levels at which community engagement may occur. At the most basic level community participation takes place in the data collection stage of a study as the researcher 'empowers' participants through their involvement in the data collection process (e.g., as interview or photovoice participants) which frames individuals not as research subjects but as providers of knowledge (Kesby, 2004). In this instance, community members have little say over the direction of the research and the researcher will return to her or his institution

Table 6.1 Levels of community involvement in participatory research

Mode of Participation	*Involvement of Local People*	*Relationship of Research and Action to Local People*
Cooption	Token; representatives are chosen, but no real input or power	On
Compliance	Tasks are assigned, with incentives; outsiders decide agenda and direct the process	For
Consultation	Local opinions asked, outsiders analyse and decide on a course of action	For/with
Cooperation	Local people work together with outsiders to determine priorities, responsibility remains with outsiders for directing the process	With
Co-learning	Local people and outsiders share their knowledge, to create new understanding, and work together to form action plans, with outsider facilitation	With / by
Collective Action	Local people set their own agenda and mobilized to carry it out in the absence of outside initiators and facilitators	By

(Cornwall, 1996)

and role as an academic commentator. Jumping a stage, Cornwall (1996) defines research where community members provide input into the research problem as being carried out at the 'consultation' level. In this instance, researchers are acting on the knowledge of the community members but are still working relatively autonomously to interpret the realities of the community.

Undertaking research in order to build community capacity requires community engagement at Cornwall's (1996) 'cooperation' level or the more intense stages of co-learning and collective action; however, a number of authors warn of the difficulty of achieving 'purist' community engagement, particularly amongst the most marginalized groups where self-determination at the collective level may be limited (Maguire, 1987; De Konig and Martin, 1996). Co-learning, for example, seeks to engage community members to determine a study focus, develop action plans and contribute to new knowledge alongside the researchers. This approach values community members as potential researchers who, by gaining tools and experience, may be equipped to carry out research at the collective action level in the future and therefore achieve social change independently. Finally, at the most engaged end of the spectrum, collective action entails the community making research decisions independent of outside facilitators. In this instance, community members may even employ researchers as consultants to address the tasks they define. The scale at which research is undertaken often influences the level of community-engagement adopted. Large-scale research is more likely to include skilled community members in the data collection or as advisors on the project (see Fenton, et al., 2002; Morisky, et al., 2004). These projects tend to involve prescribed research goals implemented in a top-down manner through strategies such as employing community interviewers. Projects involving purely lay community members, on the other hand, tend to be smaller in scale, involving fewer community members and with more modest goals (e.g., Low, et al., 2000; Maguire, 1987). These different approaches to engagement are important in shaping the potential for community empowerment.

The researcher's task, in building community capacity is to work with community members to support the development of their knowledge, skills, and resources to gain greater control over the research process. To optimize learning, Hall (1992) sees the need for researchers to use their power to strive toward assigning as much of it back to the community as possible. Thus, community-engaged research often overrides assumptions that to carry out research one must have the appropriate academic qualifications and institutional affiliations. Community-engaged research seeks to build the skills and abilities of a group to address their own health problems by supporting social action and enhancing leadership, developing new skills and building social cohesion within a community (Speer, et al., 2001). Research into civic engagement suggests that the process of fostering social cohesion may be just as important as skills gained throughout the research process. Indeed, we know that those individuals who participate civically are more likely to hold positive views about their community and share similar attitudes with other residents, whether they be political or communitarian (Lovell, et al., 2015; Speer, et al., 2001).

Critical issues in community-engaged research

At the beginning of this chapter we introduced a community-engaged research initiative that sought to address experiences of disability brought about by the accessible transit service. In this final section, we draw connections between that research experience and the wider debates around community-engaged research. Frequently, the success of participatory research is hindered by a lack of public interest in getting deeply involved in academic research (McFarlane and Hansen, 2008; Pain, 2004), the power-laden nature of research (Wilton, 2004) or an inability of the researcher to bridge the barriers to participation (Kitchin, 2001; Pain and Francis, 2003). This project was no exception and attrition of those who lost interest in the project meant that only four participants remained actively involved at the completion of the study and were interviewed for the project evaluation. We address the reasons for this and other challenges of CBPR below.

Securing community buy-in and sustaining engagement

Establishing a participatory research project involves securing buy-in support and commitment from both gatekeepers and the community. Following several unfruitful attempts at establishing a participatory research project, we approached an age-integrated social housing setting where residents included seniors and a small number of younger adults living with disabilities. With the support of the board of directors, I (Sarah Lovell) attended a meeting of the residents and introduced myself and the idea of CBPR. The residents understood the concept immediately and began to volunteer project topics before the nature of the study had even been fully explained. At the forefront of their concerns was the provision of the local parallel transit service. A week later I returned with flyers and chatted informally with residents; by that stage the project was already being described by residents as the parallel transit project. Establishing this early, personal connection with residents appeared to facilitate attendance at the first project meeting to which seven people attended. We refer to these individuals as the project 'stakeholders'.

Among the original seven stakeholders, roughly half experienced physical disabilities and used wheelchairs; the remaining stakeholders were able-bodied. The decision to focus our research on the accessible transit service was decided collectively by the group who were concerned at the long waits and limited services they experienced personally or witnessed their peers' experience. As time progressed, the able-bodied stakeholders gradually stopped attending meetings and new members who relied heavily on the bus service joined, with only two of the original group members continuing their involvement to the project's completion. The decline in stakeholder numbers placed a heavier burden on the four individuals who continued with the project and meant absences were felt more acutely. The stakeholders who remained experienced multiple chronic health conditions that affected their everyday lives in significant ways. Frequent doctor's visits, hospitalization, isolation, compromised self-care ability and ability to live independently affected their involvement in the project.

Sustaining engagement with a community requires considerable work and draws on skill sets like community organizing and facilitation which differ from those acquired through academic studies. Large-scale projects often involve community organizations where the stability of paid positions, institutional memory, and stronger formal networks serve to maintain relationships. In contrast, small, closed social networks can compromise the durability of a group, particularly where high rates of turnover are a problem. The area of critical disability studies has been at the forefront of 'enabling' forms of research in which small groups of individuals living with disabilities are active. McFarlane and Hansen (2008), for example, are two disabled activists-scholars who assembled advisory committees of disabled community members to assist in the design of their research into employment and reproductive concerns. The two researchers carried out interviews and fostered a reciprocal relationship with stakeholders, yet found their participants had little interest in continuing their involvement into academic activities such as the publication of papers in peer-reviewed journals. Likewise, Kitchin (2001) reported both inability and unwillingness to commit to a research process among the persons with disabilities he sought to engage. A different challenge arose for Wilton (2004), who warns researchers that conflicts of interest can arise when navigating between a position as an academic researcher and a disability activist (and thus also a group member). Wilton (2004, p. 120) argues that activist researchers need to allow sufficient space between themselves and the community as a means of diffusing any potential power discrepancies: 'For disability activists, researchers must be responsive to the concerns of the group with which they work, but they must keep their distance to avoid undue influence and impact on those groups.' These observations are supported by the broader participatory research literature which emphasizes the need for the researcher to step back from a leadership role and play a facilitator role in order for community empowerment to be meaningful (Tengland, 2006). Our challenges achieving community engagement in participatory research led to further research into the factors that aid community capacity building. Importantly, communities can range widely in their readiness for action, but fostering community trust and identifying the community member who makes things happen – often not the leader – is instrumental to the success of outsider interventions to build community capacity (Lovell, et al., 2011).

Community members as researchers

Participatory researchers have reinterpreted who is qualified to carry out research, questioned the structures of academia, and sought to overcome experiences of oppression in society. The work of educator Paulo Freire (1970) is fundamental to understanding the ideology underpinning this approach. Freire (1970) conceives of knowledge as something that should not be contained within ivory towers, but rather embedded in the social realities of those who live through oppression. As such, participatory researchers often reject the values ingrained in much of positivist science and work instead toward an understanding of social life that is embedded in the critical research traditions that precede it, particularly feminist and radical perspectives.

Ideologies of community-engaged research emphasize the benefits for the community, but alongside these benefits come the challenges of sustaining community engagement and negotiating the role of the researcher in the process. The collaborative nature of community-engaged research creates many incompatibilities with presiding academic structures. The researcher, in favoring community needs over the achievement of academic markers of success, does not have control over the direction that the research will follow (Heaney, 1993).

Researchers have argued that the opportunities we offer communities for participation may be more significant to our outcomes than the research technique we choose to adopt, particularly where the goal is to build community capacity (Cornwall and Jewkes, 1995). Having sought a deep level of participation, our experience in the disability transportation study suggests that research is just one, potentially minor component of CBPR for stakeholders. The wider experience of transformative research encompasses community engagement, research and transformative action. We had sought to support the stakeholders as decision-makers, data collectors and active participants in the dissemination of research results. Early on, community stakeholders played a key role in decision-making by opting to undertake focus groups to investigate the experiences of accessible transit users and to elicit opinions on the strengths and weaknesses of the service. Focus groups are a means of gaining rich data and gauging collective opinions while knowledge is accumulated through the conversation of participants. Stakeholders anticipated that the social element of a focus group would be valued by the disabled participants both as a source of support and a means of 'giving back' to participants. While stakeholders were keen to facilitate focus groups and we undertook training to support them in this role, attendance proved challenging. Ultimately, stakeholders attended and facilitated only two of six focus groups due, ironically, to the unavailability of the accessible transit service and health and chronic conditions, as Lianne explains:

> No, I would have liked to have someone to go to these [focus group] meetings . . . so you didn't have to do it all yourself, but health-wise, as I say, that's one of the things that I'm putting myself on a regime because I never know and like I might go down and nothing happens, but I also know myself and I also know that within a month I can go down with the blink of an eye and then, you know, you feel bad because you're invited to go down and you just don't hold up your end and there's not a thing you can do about it. And it does, it makes you feel bad because I gave my word . . . I gave my word and I feel that I should be able to uphold my word. So, I find that difficult.

While stakeholders were keen to be involved in the focus groups, full participation throughout the project was quickly seen as both infeasible and undesirable by the stakeholders. Subsequently, Sarah Lovell played a major role in organizing the project, managing the research process and sustaining momentum with the stakeholders. Perhaps unsurprisingly, stakeholders were interested in those components of the research that relied on less academic skills (the facilitation of focus groups, the planning and decision-making surrounding the project, and the analysis of the research data) rather than pursuing ethical applications or writing up what they learned either to present to the local or as co-authors of academic research. Opportunities for challenging existing power structures, strengthening public speaking skills, and collective group action that were particularly valued as stakeholders sought to ensure the group gained a public voice. While ideologies around community-engaged research value intense community involvement throughout a project, our experiences suggest the strategic involvement of stakeholders may be just as effective in empowering communities, while minimizing participant burden.

All researchers are driven to meet their own political ends, whether they are academics seeking publishable results or participants wishing to expose and challenge the presence of inequalities within their communities. This recognition can create tensions with the common perception that scientists hold objective and impartial knowledge. Participatory research upsets this balance by arguing that those invested in the problem can be equally knowledgeable. So, do these underlying political interests have implications that extend beyond academia? If we are to heed Wallerstein's (2004, p. 10) words, the answer would be, 'yes'. The disinterestedness and the qualifications of the researcher are central factors in the public's ability to place trust in the scientific community:

> We assume that specialized knowledge is difficult to acquire, demanding long and rigorous apprenticeship. We put our faith in formal institutions, which in turn are evaluated by reliability scales. . . . In short, we trust that professionals have appropriate skills, and most particularly the skill to evaluate new truth claims in their fields.

If we encourage people in the community who have no formal academic training or credentials to carry out research, are we undermining trust in research as an institution and thus future opportunities for ourselves to carry out research? The answer to this question likely lies in the research perspective we subscribe to and our commitment to participatory approaches.

Building community capacity through research

Community-engaged research in health geography has largely flown under the radar (e.g., Kitchin, 2001; Wilton, 2004). There is a significant gap in our knowledge of how engagement is best achieved with populations who have compromised health or are living with disabilities. This chapter posits that participatory research must extend beyond the application of participatory methods within traditional research frameworks and instead engage with broader social goals (e.g., social justice). This will require the questioning of power structures, or we risk community-engaged research becoming merely another research tool for accessing the views of community members without any engagement with the goals and aspirations of the community members. In other studies, community empowerment has been found to take in excess of seven years to achieve and is evident in experiences of successful lobbying for changes to government policy (Raeburn, 1993, as cited in Laverack and Wallerstein, 2001). Over the course of the disability transit project, we attended the annual general meeting of the accessible transit service, raised awareness of the shared issue within the disability community and wrote up a report of our focus groups with bus users which we circulated to the accessible transit services board. The most surprising finding on evaluating the stakeholder's experiences was that two of the four believed positive change resulted from the project through improved accessible transit services. They, however, noted the possibility that they may have been receiving privileged treatment as a result of their united voices. The stakeholders also recognized the potential for further, positive change as we worked with the study results and pursued further disability related issues.

Lianne: When [June] came home yesterday from the hospital and checked her answering machine, [an accessible transit service employee] was on the answering machine and she profusely apologized to her for the bus being late and she hoped she wasn't late for her appointment and the reason the bus wasn't there was because it was involved in an accident. And I said to her, I said well boy, we must have made a difference because – they never would have done that.

Another stakeholder, who joined the project later and relied heavily on the parallel transit service was less critical of the service and did not see the same transformative potential for the group. Rather, she indicated that much of the problems accessible transit users experienced could be resolved with better education regarding service provision and suggested the impetus for action rested with the accessible transit service rather than the users themselves. The stakeholders who identified our presence as having produced positive change were equally quick to point out the difficulties of achieving wider change, particularly in the face of strong political forces. However, finding that the accessible transit service had improved in terms of both availability and customer service was a major boost for the stakeholders and confidence in what we could achieve grew as a result.

June: It would be nice if the people at the [accessible transit service] would realize that we know what we're talking about . . . the problems that they have and trying to do something about fixing them but, like I said, the service is getting a little bit better so maybe they've got wind of what's going on and maybe they're trying. That would be nice.

Lianne: I think we did well. I really do – We at least tried, you know. Like, don't sit back and complain about things if you're not willing to try to make a difference and we did at least try so. . . . And, as you say,

> it's not over yet. Maybe somebody'll turn around in mid-stream and really make up their mind that they want to do something about it.

While positive change may have been an outcome of the project, our collective efforts were seen to have little impact on wider community capacity. Stakeholders indicated that there was a high level of agreement and general cohesion among the group, but small numbers and a lack of individuals with a high level of initiative or leadership were seen to hinder the process. The greatest frustration was directed toward those who had left the project or chosen not to participate.

In developing the proposed research we had hoped to achieve community engagement at the co-learning level (see Table 6.1). Accordingly, decisions were to be made primarily by the stakeholders, and the researcher was to be a source of research skills and facilitation. However, the level of participation achieved was much closer to a relationship of cooperation where priorities were determined by the community stakeholders, and the responsibility for the research process remained with the researcher. While community-engagement in research is intended to minimize power imbalances between the researcher and participants, in this case stakeholders were reliant on the researcher's information and advice to make methodological decisions, and this reliance widened the gap between the researcher and the stakeholders. While we sought to emphasize their role as disability experts, the experience contrasted with participatory research literature, which idealizes community engagement and argues that for participation to be meaningful stakeholders must be involved in all aspects of the research and learn skills that would enable them to repeat the process in the absence of the researcher (Cornwall, 1996). Toward the end of the project the stakeholders began to look forward to future problems, but undertaking research themselves did not appear to be a part of that plan. This was most evident when the researcher showed up to a meeting and was told about their plans for her to look into the accessibility of shops once the current project was complete: 'We've got the next problem lined up for you'. Evident in this discourse were the stakeholders' issues, or 'their' problems, becoming 'our' project, a reflection of the modest level of participation achieved through the day-to-day research activities.

The stakeholders saw research as a valuable tool yet preferred to be spokespersons and idea generators rather than co-investigators undertaking research. Living with a disability or chronic condition had led many of the stakeholders to develop strategies for managing their energy levels by limiting their day-to-day commitments, and this practice extended to the CBPR project. Stakeholders took on roles that allowed them to voice their concern and discuss with their peers the nature of the problem without becoming heavily involved in the time consuming details and practicalities of research.

Conclusions

A loose précis for the conduct of research to build community capacity has emerged from the collective works of Freire (1970), and Minkler and Hancock (2003) that identifies public ownership over knowledge, a critical awareness of the root causes of major social issues, and community control over the search for solutions as fundamental to individual and community empowerment. Participatory researchers have sought to increase the transparency and accessibility of research to persons in the community who have neither formal academic training nor credentials as a means of enhancing its relevance to the public and strengthening community control over outcomes. But as our experience illustrates, academic research may be viewed by the community as merely a tool in achieving a political voice and of less relevance to their daily lives. Perhaps, we are conflating the power of research with our power to be heard as academics?

Health geographers have a long history of undertaking community-engaged research and are increasingly bringing these experiences to the attention of the scholarly community. In this work, health geographers are highlighting the tensions and the achievements of participatory methods, activist encounters and more traditional research collaborations that are transforming both communities and researchers. This chapter has highlighted the range of approaches to community-engaged research and the difficulties of building community capacity through collaborative approaches.

References

Andrews, G. J., Evans, J., Dunn, J. R. and Masuda, J. R. 2012. Arguments in health geography: on sub-disciplinary progress, observation, translation. *Geography Compass*, 6(6), pp. 351–383.

Askins, K., 2009. 'That's just what I do': placing emotion in academic activism. *Emotion, Space and Society*, 2, pp. 4–13.

Bennett, D., 1991. Explanations in medical geography: evidence and epistemology. *Social Science and Medicine*, 33(4), pp. 339–46.

Bijoux, D., and Myers, J., 2006. Interviews, solicited diaries and photography: 'new' ways of accessing everyday experiences of health. *Graduate Journal of Asia-Pacific Studies*, 4(1), pp. 44–64.

Brown, T., and Duncan, C., 2002. Placing geographies of public health. *Area*, 34(4), pp. 361–369.

Castleden, H., Garvin, T., and First Nation, H.-A.-A. 2008. Modifying photovoice for community-based participatory Indigenous research. *Social Science and Medicine*, 66(6), pp. 1393–1405.

Cornwall, A., 1996. Towards participatory practice: participatory rural appraisal and the participatory process. In De Konig, K., and Martin, M. (eds.) *Participatory research in health: issues and experiences.* London: National Progressive Primary Health Care Network.

Cornwall, A., and Rachel J., 1995. What is participatory research? *Social Science and Medicine*, 41(12): pp. 1667–1676.

Darlaston-Jones, D., 2007. Making connections: thc relationship between epistemology and research methods. *The Australian Community Psychologist*, 19(1), pp. 19–27.

Davidson, J., 2001. 'Joking apart . . . ': a 'processual' approach to researching self-help groups. *Social and Cultural Geography*, 2(2), pp. 163–183.

De Konig, K., and Martin, M., 1996. *Participatory research in health: issues and experiences.* London: National Progressive Primary Health Care Network.

Dorling, D., 2014. Inequality and the 1%. Brooklyn, NY: Verso Books.

Fenton, F., Chinouya, M., Davidson, O., and Copas, A., 2002. HIV testing and high risk sexual behaviour among london's migrant african communities: a participatory research study. *Sexually Transmitted Infections*, 78, pp. 241–245.

Freire, P., 1970. *Pedagogy of the oppressed.* New York: Center for the Development of Social Change.

Gibson-Graham, J. K. 1994., 'Stuffed if I know!': reflections on post-modern feminist social research. *Gender, Place and Culture*, 1(2), pp. 205–224.

Gilbert, M., and Masucci, M., 2008. Reflections on a feminist collaboration: goals, methods and outcomes. In Moss, P., and Al-Hindi, K. (eds.) Lanham, MD: Rowman and Littlefield.

Green, L., and Mercer, S., 2001. Community based participatory research: can public health researchers and agencies reconcile the push from funding bodies and the pull from communities. *American Journal of Public Health*, 91(12), pp. 1926–1943.

Hall, B., 1992. From margins to center? The development and purpose of participatory research. *American Sociologist*, 23(4), pp. 15–28.

Hay, I., 2000. *Qualitative research methods in human geography*, Melbourne, Oxford University Press.

Heaney, T., 1993. If you can't beat 'em join 'em: the professionalisation of participatory research. In Park, P., Brydon-Miller, M., Hall, B., and Jackson, T. (eds.) *Voices of change: participatory research in the United States and Canada.* Toronto: OISE Press.

Kearns, R., and Moon, G., 2002. From medical to health geography: novelty, place and theory after a decade of change. *Progress in Human Geography*, 51(7), pp. 1047–1060.

Kearns, R. A., Collins, D. C. A., and Neuwelt, P. M., 2003. The walking school bus: extending children's geographies? *Area*, 35(3), pp. 285–292.

Kesby, M., 2000. Participatory diagramming: deploying qualitative methods through an action research epistemology. *Area*, 32(4), pp. 423–35.

Kesby, M., 2004. Participatory diagramming and the ethical and practical challenges of helping Africans themselves to move HIV work "beyond epidemiology." In Kalipeni, E., Craddock, S., Oppong, J., and J. Ghosh (eds.) *AIDS in Africa: beyond epidemiology.* Melding: Blackwell.

Kindon, S., Pain, R., and Kesby, M., 2005. *Participatory action research: connecting people, participation and place*. London: Routledge.

Kitchin, R., 2001. Using participatory action research approaches in geographical studies of disability: some reflections. *Disability Studies Quarterly*, 21(4), pp. 61–69.

Labonte, R., 1997. Community, community development and the forming of authentic partnerships: some critical reflections. In Minkler, M. (ed.) *Community organizing and community building for health.* New Brunswick, NJ: Rutgers University Press.

Laverack, G., and Wallerstein, N., 2001. Measuring community empowerment: a fresh look at organizational domains. *Health Promotion International*, 16(2), pp.179–185.

Lovell, S. A., Gray, A. R., and Boucher, S. E., 2015. Developing and validating a measure of community capacity: Why volunteers make the best neighbours. *Social Science and Medicine*, 133, pp. 261–268.

Lovell, S. A., Kearns, R. A., and Rosenberg, M. W., 2011. Community capacity building in practice: constructing its meaning and relevance to health promoters. *Health and Social Care in the Community*, 19(5), pp. 531–540.

Low, J., Shelley, J., and O'Connor, M., 2000. Problematic success: an account of top-down participatory action research with women with multiple sclerosis. *Field Methods*, 12(1), pp. 29–48.

Maguire, P., 1987. *Doing participatory research: a feminist approach.* Amherst, MA: The Center for International Education.
McDowell, L., 1992. Doing gender: feminism, feminists and research methods in human geography. *Transations of the Institute of British Geographers*, 17(4), pp. 399–416.
McFarlane, H., and Hansen, N., 2008. Inclusive methodologies: disabled people in participatory action research in Scotland and Canada. In Kindon, S., Pain, R., and Kesby, M. (eds.) *Participatory action research: connecting people, participation and place.* London: Routledge.
Minkler, M., and Hancock, T., 2003. Community-driven asset identification and issue selection. In Minkler, M., and Hancock, T. (eds.) *Community-based participatory research for health.* San Francisco, CA: Jossey-Bass. pp. 135–154.
Mistry, J., Bignante, E., and Berardi, A., 2014. Why are we doing it? Exploring participant motivations within a participatory video project. *Area*, p. Early View.
Morisky, D., Ang, A., Coly, A., and Tglao, T., 2004. A Model HIV/AIDS risk reduction program in the Philippines: a comprehensive community-based approach through participatory action research. *Health Promotion International*, 19(1), pp. 69–76.
Myers, J., 2010. Moving methods: constructing emotionally poignant geographies of HIV in Auckland, New Zealand. *Area*, 42(3), pp. 328–338.
Oliver, M., Witten, K., Kearns, R., Mavoa, S., Badland, H., Carroll, P., Drumheller, C., Tavae, N., Asiasiga, L., Jelley, S., Kaiwai, H., Opit, S., Lin, E.-Y., Sweetsur, P., Barnes, H., Mason, N., and Ergler, C., 2011. Kids in the city study: research design and methodology. *BMC Public Health*, 11(1), pp. 1–12.
Ortega-Alcazar, I., and Dyck, I., 2011. Migrant narratives of health and well-being: challenging 'othering' processes through photo-elicitation interviews. *Critical Social Policy*, 32(1), pp. 106–25.
Pain, R., 2004. Social geography: participatory research. *Progress in Human Geography*, 28(5), pp. 652–63.
Pain, R., and Francis, P., 2003. Reflections on participatory research. *Area*, 35(1), pp. 46–54.
Park, P., 1993. What is participatory research? A theoretical and methodological perspective. In Park, P., Brydon-Miller, M., Hall, B., and Jackson, T. (eds.) *Voices of change: participatory research in the United States and Canada.* Toronto: OISE Press.
Potter, G., 1999. *The philosophy of social science: new perspectives.* Abingdon, Oxon: Prentice Hall.
Richmond, C., and Fortier, J. (directors), 2013. *Gifts from the Elders* (film). London, Ontario: University of Western Ontario.
Rosenberg, M., 2013. Health geography I: social justice, idealist theory, health and health care. *Progress in Human Geography*, 38(3), pp.466–475.
Routledge, P., 1996. The third space as critical engagement. *Antipode*, 28(4), pp. 399–419.
Speer, P., Jackson, C., and Petersen, A., 2001. The relationship between social cohesion and empowerment: support and new implications for theory. *Health Education and Behavior*, 28, pp. 716–732.
Taylor, M., 2007. Community participation in the real world: opportunities and pitfalls in new governance spaces. *Urban Studies*, 44(2), pp. 297–317.
Taylor, M., 2014. 'Being useful' after the ivory tower: combining research and activism with the Brixton Pound. *Area*, 46(3), pp. 305–312.
Tengland, P., 2006. Empowerment: a goal or a means of health promotion? *Medicine, Health Care and Philosophy*, 10, pp. 197–207.
Thomas, F., 2007. Eliciting emotions in HIV/AIDS research: a diary-based approach. *Area*, 39(1), pp. 74–82.

Tobias, J., Richmond, C., and Luginaah, I., 2013. Community-based participatory research (CBPR) with indigenous communities: producing respectful and reciprocal research. *Journal of Emprical Research on Human Research Ethics*, 8(2), pp. 129–40.

Walker, S., Eketone, A., and Gibbs, A., 2006. An exploration of kaupapa Maori research, its principles, processes and applications. *International Journal of Social Research Methodology*, 9(4), pp. 331–344.

Wallerstein, I., 2004. *The uncertainties of knowledge*. Philadelphia: Temple University Press.

Wilton, R., 2004. Keeping your distance: balancing political engagement and scientific autonomy with a psychiatric consumer/survivor group. In Fuller, D., and Kitchin, R. (eds.) *Radical theory/critical praxis: making a difference beyond the academy?* Vermon and Victoria, Canada: Praxis(e)Press.

7 Walking in their shoes

Utilizing go-along interviews to explore participant engagement with local space

Jennifer Dean

Reflections on praxis

I have used go-along interviews for several research studies in order to explore the relationship between place and health, but I first tried the method as part of my doctoral research and learned the most from that early experience (Asanin Dean, 2012). The doctoral study examined local factors that influence adolescent body weight in low-socioeconomic neighborhoods, and I was particularly interested in linking perceptual data from adolescents with data from key informants about the various healthy eating and physical activity options that existed, along with ecological data of the neighborhoods.

While the benefits of go-along interviews are plentiful and discussed later in the chapter, I encountered several challenges during my first (and subsequent) experience using the method. The first challenge occurred prior to entering the field. The institution's research ethics board (REB) had strong reservations about the safety and well-being of participants while the interview was taking place. The REB worried that I would be intimidating to high-risk youth, and that I would stick out as an 'outsider' in their community, thus compromising confidentiality. Specifically, they were concerned that my presence, along with the audio recorder and microphone, would somehow mark the youth as 'rats', potentially outing them and the researcher as undercover informants reporting on the ills of the community. I was stunned by what I perceived to be stereotyping by members of the board. (Do high-risk youth *look* significantly different from graduate students? Would the concerns about informants reporting on deviant behavior remain if the research was in a higher income neighborhood?) And I was further concerned over whether my research, meant to improve policy for the design of healthier built environments, would be able to proceed as planned.

The concerns of the REB were eventually alleviated after the board requested to see me in person and agreed that I was neither intimidating nor did I have the appearance of an undercover police officer who would potentially 'out' the youth or community members. I was both relieved and offended that my appearance and demeanor were interrogated by the board prior to being granted approval. As a graduate student I felt powerless to comment about this process formally, but I wondered how many other researchers who worked with marginalized populations had been called to the board for a visual assessment.

At that meeting I had significant discussion with the board about how the go-along would affect the safety of participants. The board's underlying assumption that low-socioeconomic status neighborhoods were inherently dangerous meant that walking around those same neighborhoods would be risky for the youth. I made a strong case for the fact that youth were simply walking in their *own* neighborhood during daylight hours and along routes that they have travelled many times before. Therefore, I argued, the risks were typical of what they would experience on a daily basis (even if it was in a neighborhood with higher than average crime rates, the risks of participating were no greater than what they would encounter during their daily commute to school). The board countered with suggestions to do an imagined 'virtual' neighborhood walk, or use an existing map to trace travel routes rather than go in person, but that type of cognitive mapping was different from the multisensory experience of go-alongs. Approval was granted for the go-alongs but under the conditions that the virtual neighborhood walk would be an option for those participants who felt unsafe walking around with me (this was never a concern for participants in this study) and that participants and their parents knew exactly what they were getting into (informed consent) when agreeing to participate in the go-along. To achieve this, we created a brief recruitment video for our study website that detailed the process of the go-along interview and showed an example of two adolescents (family members of the research team) engaging in a (mock) go-along interview with me. Ironically this video was posted on the REB website as an 'exemplar' of informed consent, though it did not seem to be overly effective with regards to recruitment. When I have proposed to use go-along interviews for other studies and applied to different REBs, the concerns for participant safety (e.g., getting robbed for their GPS watch) and confidentiality (e.g., wearing a microphone as a marker of research participation) are the primary issues raised, but have not been problems in the field. I have yet to have a participant decline a go-along interview due to concerns over neighborhood safety; this may be due in part to sampling bias – those concerned with safety do not proceed past the letter of information – but also because I allow participants to choose the time of day for the interview, and they have control over where in their neighborhood we go. With respect to be marked as a research participant, most find the microphone to be discrete (but see below for technical issues), and when encountering neighbors during the interview, some have even introduced me as a researcher and briefly discussed the research, which has acted as real-time snowball sampling, resulting in two additional participants in one study. In my experience, these far-reaching concerns raised by the REBs are in part because the method is still relatively new and unfamiliar to many board members. It is also symptomatic of what an emerging body of literature has described as 'ethics creep' and the overreaching of social science and humanities REBs in particular (see Dyer and Demeritt, 2009; Haggerty, 2004; Juritzen, Grimen, and Heggen, 2011).

A second challenge in employing go-alongs was participant buy-in. While many youth in this study – and adults in other studies using go-along interviews – were keen to show off their neighborhood, others argued that inclement weather (rain, snow, cold, heat, fog) made the interview process undesirable. In some cases, the go-along component ended up being optional, and in certain circumstances (i.e., bad weather, too tired) the participants preferred to stay indoors. In studies where the go-along was not optional, participants seemed more willing to take part or would request to reschedule when given the option.

One thing rarely talked about in research training is the technical issues related audio recording devices. I commonly use a lapel microphone that discretely fits on to a jacket or scarf connecting to the audio recorder that fits into that participant's pocket (the cord can run inside the jacket for more discreet practice). In addition to the usual technical difficulties with recording, this approach also resulted in poor sound quality due to a muffled microphone from the participant's clothing or scarf, or due to wind or heavy traffic. Further, since the audio recorder is in the participant's pocket, it is difficult for the interviewer to monitor for low battery and accidental (or intentional) turn off. It is possible, with higher quality recorders, for the researcher to wear the microphone and recorder, or to use props (e.g., a coffee cup) to hide the microphone in order to mitigate some of these technical issues.

A final challenge is related to researcher safety especially given that some neighborhoods were unfamiliar to the interviewer. This makes navigation difficult, especially as you follow another's lead. Moreover, in the case of youth, neighborhood infrastructure was used in unconventional ways, which I affectionately termed 'off-roading' (i.e., walking on curbs, going under broken fences, going down back alleys, through peoples' backyards [friends and family of the participants], crossing busy roads against or without the lights) that often was out of the my own comfort zone as a pedestrian in an unfamiliar area. I often kept a printed map of the community in my back pocket, a cellphone, and had someone who knew where I was and when I would be back during each interview. Ironically, researcher safety has not been a concern of any of the REBs I have worked with, though supervisors and departments often have their own safety standards to be adhered to.

Like all research strategies, the go-along interview method is not without its challenges but it is an exciting and innovative method that I have continued to employ since my doctoral research project. The remainder of the chapter will detail the logistics of go-along interviews based on the literature and my own experiences, having completed one study using go-along interviews and being in the process of completing several more. I will also spend significant time discussing how this method can contribute new insights into emerging theories and topics in health geography.

Introduction

Go-along interviews are still relatively new as a qualitative research method in the social and health sciences, but its roots are familiar. The method combines qualitative interviewing, participant observation and field observations as a way to better understand how environments shape human behavior, and how residents use, interpret and experience local environments (Kusenbach, 2003; Carpiano, 2009).

My own research has used go-alongs with various populations for differing objectives. For example, in one study I incorporated go-along interviews to explore how residential neighborhoods influence adolescent physical activity and diet (Asanin Dean, 2012). In another study go-alongs are combined with a community mapping exercise to examine the impacts of green space on recreational walking routes (ongoing). An additional study I am involved in uses go-alongs to understand subjective measures of walkability among seniors (ongoing). Throughout the chapter, examples of others who have used go-along interview will be provided alongside details of how and when to use this method.

Understanding the go-along interview method

Go-along interviews are conducted in situ as participants provide the interviewer a tour of their neighborhood (Carpiano, 2009) and can rely on different modes of transport: walking or riding (bicycles, public transportation, private vehicles), which have been similarly termed 'ride-alongs'. Kusenbach (2003) identified five thematic areas that are easily explored using go-along interviews:

1. *Perceptions* – to understand how personal values guide experiences of the social and physical environment on a daily basis,
2. *Spatial practices* – to examine the various degrees and forms of engagement with the environment,
3. *Biographies* – to explore the relationships between life histories and places, and therefore the meaning behind mundane life routines,
4. *Social architecture* – to assess the types of relationships between people and how they orient themselves within a social setting,
5. *Social realms* – to investigate the patterns of interaction and how place shapes this.

The selection of approaches used when employing this method will depend on the research objectives. While many of the benefits and challenges of go-alongs naturally align with those of more traditional interviews, participant observation and field observation approaches (Patton, 2002), some of the more unique strategies, benefits and challenges will be discussed here.

As an interview strategy, researchers are literally 'walked through' the lived experiences of a particular area while the constant engagement with place allows participants to remember or rediscover new elements of their environment as they pass physical prompts or environmental cues (Carpiano, 2009). Eisenberg and colleagues (2012) examined students' knowledge of campus health services using go-along interviews and found that students were able to recall new services and remember certain health services only when seeing things physically in front of them. The go-along approach encouraged and reminded students where known resources were located, and in addition, many students would actively seek out or accidentally discover resources not originally known to them (Eisenberg, et al., 2012). As such, the environmental landmarks themselves are similar to and sometimes better at eliciting more information from participants than verbal prompts and probes commonly used in interviews (Eisenberg, et al., 2012). Environmental cues can also be helpful for interviewers for prompting additional questions. Gardner (2011) suggests looking for measures of *erosion* (degree of wear) and *accretion* (material build up) that may be used to identify important places. For example, footprints across an unpaved portion of a park may suggest a frequent travel route, or a store with newspapers or mail piled up may indicate its closure, which can prompt further questions.

The mechanics of go-alongs are similar to qualitative interviews in that they can be semi-structured or a more open narrative approach, though the latter may prove to be more accommodating given the unplanned situations that may occur in the field (e.g., seeing a friend, witnessing a car accident or dispute) (Kusenbach, 2003). Yet those unplanned situations themselves can be a goldmine of information about how the individual interacts with both social and physical space. Similarly, the silences that can be awkward in a face-to-face sit down interview (though an important tool for researchers), can be more comfortable and helpful during a go-along interview because it allows space for thinking about new ideas triggered by surroundings (Bergeron, et al., 2014). Accordingly there are several lines of questioning that are especially helpful during go-along interviews including: why participants chose a particular path, whether the participant feels comfortable or confident in the environment, what comes to mind when walking in an area, what their typical experience of that environment is like, what places evoke special meanings, emotions or concerns (Kusenbach, 2003; Carpiano, 2009; Bergeron, et al., 2014).

To illustrate with an example, in my doctoral research on environmental determinants of adolescent body weight, part of the protocol involved asking adolescents to give me a tour of their neighborhood while I asked questions. This go-along portion of the interviews lasted between 20 to 75 minutes and varied in distance (both by urban and suburban neighborhood status, and by participants' familiarity with the neighborhood). Participants put on a small lapel microphone that was plugged into an audio recorder kept in their pocket and I had the script memorized so that I could walk freely with them. During the walk, I would ask participants about whether they used a park or plaza we had just passed, if they always took a particular route to school, or what existed down a particular path or road we had passed. I also wanted to know whether the physical infrastructure promoted by planners and public health specialists was actually used by youth and found that in many cases sociocultural and political factors limited how youth used their neighborhood (Dean and Elliott, 2012). For instance, one neighborhood had outdoor parks retrofitted with fitness equipment, yet most youth in that area said they did not have any parks they could play at. During the go-along interview I would ask them about the park as we passed, and they shared that they did not use it because it was considered part of gang territory and had been the site of a recent stabbing. One participant's parents forbade her from walking down a street in her neighborhood (which would have streamlined her route to school) because a known drug dealer lived there. These were important insights into the neighborhood gathered by experiencing it with the participant, and more practitioners are using guided neighborhood walks to gain experiential and perceptual information from residents (see Hamilton, 2015; Jane's Walk, 2015).

In terms of limitations and challenges, taking notes can be difficult and, in some cases, inappropriate during the interview, so audio recording is especially essential, as is creating field notes immediately after the interview while paying particular attention to the locations and landmarks mentioned or visited during the interview (Kusenback, 2003; Garcia, et al., 2012). It has been argued that the level of activity and the novelty of go-along interviews can be exciting for participants, which may result in potentially higher rates of participation (Garcia, et al., 2012), yet conversely, go-along interviews may be unappealing to individuals with limited mobility or serious concerns about safety in their community (Carpiano, 2009; Asanin Dean, 2012). There are mixed findings on the role of weather in determining a participant's willingness to take part in a go-along interview (Carpiano, 2009; Asanin Dean, 2012; Garcia, et al., 2012), with extreme heat, cold and precipitation being general turn-offs.

In terms of participant observation, go-along interviews provide insight into the physical environment and how people perceive, process, and navigate an area (Carpiano, 2009). Kusenbach (2003) suggests that go-along interviews are a systematic version of 'hanging out' with participants, while Carpiano (2009) notes that during go-along interviews the researcher is taken on a 'spatialized journey' and learns about an area through the participant and his or her own experiences within it. To encourage accurate and authentic observations, it is important to conduct the go-along in a way that follows participants in their most familiar environments and as closely as they normal engage in space possible (e.g., performing the interview on a particular day of the week, time of day, and on a specific route) (Kusenbach, 2003), though others have suggested that true spontaneity and 'natural' observations are difficult to achieve in any form of overt participant observation, including go-along interviews (Bergeron, et al., 2014). Researchers have mitigated this issue differently in the past, dependent on their overall aim. On one hand, Carpiano (2009) describes sitting down with participants prior to the go-along to decide on a route, which is a strategy I have used to understand the decisions a participant makes to travel a particular route between two frequented locations and not others. To explore more deeply, if the travel decision is determined by time, distance, function, community aesthetics, or quality of the physical infrastructure. Bergeron and colleagues (2014) on the other hand, advocate for minimal guidance and direction from researchers in order to allow for spontaneity on the walk. This approach has proved to be helpful in my own work with adolescents when I requested a tour of the neighborhood and asked participants to treat me as a newcomer to the area. In such circumstances, I paid close attention to the landmarks they thought were most significant, I watched how they actually used and avoided the space around them, and was able to travel with them as they went 'off-roading' as well as had my own experience of the environment. Allowing participants to choose their own route tends to shift traditional power dynamics between researcher and participant by empowering the participant to take the literal and figurative lead of the interview (Jones, et al., 2008; Carpiano, 2009). The power shift is arguably only temporary as it lasts through the data collection stage, though reflexive researchers may utilize other strategies to maintain a balance of power, such as incorporating other participatory data collection approaches (see later in the chapter), member checking, and engaging participants or community representatives in the analysis stage. In addition, go-along interviews can also be helpful in building relations between the researcher and community, especially in environments where trust and legitimacy can be called into question or access through community gatekeepers otherwise predominates.

Go-along interviews also incorporate field observation methods, which are beneficial ways for the researcher to become familiar with the study site, make observations that participants might not notice themselves, and ask questions inductively based on prompts gleaned from those observations (Kusenbach, 2003; Evans and Jones, 2011). For example, when walking with research participants during the Canadian winter, the presence of snow banks that block sidewalks or footprints on unplowed pathways, resulted in questions to participants about the adequacy of snow removal services or public works department in their neighborhood. This resulted in recommendations for prioritizing snow removal of walkways and sidewalks ahead of streets in order to accommodate or even promote highly walkable neighborhoods.

Further, go-along interviews can provide the contextual insights offered by traditional ethnographic fieldwork, but without the long-term fieldwork that is normally required (Carpiano, 2009). I have used go-along interviews as an exercise in 'ground-truthing' (Paquet, et al., 2008; Sharkey and Horel, 2008), which is a practice widely used by physical geographers to validate remote sensing data and increasingly utilized by social scientists (Paquet, et al., 2008). I used the field observation component of go-along interviews to verify what adolescents and key informants were telling me about the local environment, finding inconsistencies between the two groups and between what adolescents said about their environment and what I observed. To illustrate with a previous example, key informants noted that the neighborhood was well served by playgrounds and parks, but adolescents reported the neighborhood did not have any parks for them to play at. I interpreted this as an absence of physical infrastructure rather than what I learned to be the reality: there was park space available, but it was not usable due to criminal activity.

Researcher: This park is close to your house [pointing]. Do you ever go here?

Participant: [Grandma] doesn't normally let us come here anymore because there was an incident with a gun and a couple of people getting shot, and one woman getting run over by a car when she was running away. . . . We didn't know about these events at first,

> and then we ended up getting told that we weren't allowed to go, because of what happened. So none of us [friends and sisters] comes here anymore. (12 year old, female)

As a form of field observation, go-along interviews are similar to community safety audits (METRAC, 1989), systematic social observation (Sampson and Raudenbush 1999) and neighborhood walkability and active living audits (Kelly, et al., 2011; Bors, et al., 2012) with the added benefit of real-time subjective interpretations of the environment. It is helpful to take additional field notes of the specific locations and various landmarks from the go-along to accompany the audio recording. This is to allow for easier analysis of information when all sources of data are combined (Carpiano, 2009; Evans and Jones, 2011).

Go-along interviews can also be mixed with other methods. Carpiano (2009) advocates for the use of go-along interviews in combination with other methods, such as more traditional in-depth interviews or additional field observations. In the adolescent body weight study, I combined in-depth interviews with go-along interviews to gather additional data about adolescents' perceived determinants of body weight including and beyond their local environment, and combined this data with additional contextual data from key informants (e.g., public health specialists, city councilors, urban planners and youth workers) who were familiar with city and neighborhoods (Dean and Elliott, 2012; Asanin Dean, 2012). This mixed-methods approach allowed for data triangulation and a comprehensive understanding of the environment (i.e., what exists in the actual environment in terms of amenities, demographics, crime rate, community issues) and adolescent interaction (Farmer, et al., 2006). To cite another example, Evans and Jones (2011) compared participants who were involved in sedentary interviews to those who participated in go-along interviews within the same study. This design allowed for exploration of what participants say and where they say it, as well as the content differences between participants using the two interview strategies. The results of their study indicate that the go-along interviews were highly informed by the landscape in which they took place generating more depth in terms of place-specific data. Not surprisingly, this method was found to be less productive if the main purposes of the study were to gain autobiographical narratives (Evans and Jones, 2011).

Go-alongs have also been combined with quantitative and geospatial data collection methods. For instance, Chow (2011) conducted a survey of food allergic adults and parents of food allergic children about food purchasing behavior and the practice of reading food labels. They further explored the quantitative results by conducting go-along interviews in grocery stores with parents of food allergic children to learn more about how the context (i.e., grocery stores) influenced their purchasing behavior (e.g., reading of food labels). A more common approach combines go-along interviews with geospatial data collected through the use of GPS trackers (e.g., watches, cell phones) or by geocoding routes and specific amenities and locations after the fact (Evans and Jones, 2011; Bergeron, et al., 2014; Pawlowski, et al., 2014). The qualitative data can at times be synchronized with the GPS data, creating a qualitative geographic information system

(GIS) that connects personal narratives and experiences to the actual place in which they occur or were recalled (Evans and Jones, 2011; Bergeron, et al., 2014; Garcia, et al., 2012). For example, Bergeron and others (2014) took an integrative approach to collecting and presenting their data. While traveling, a GPS was used to monitor travel paths of each participant and to reference where participants photographed their local environment. In this study, the go-along interviews followed an unstructured format with researchers giving minimal guidance to the participants, while occasionally questioning participant's perceptions of the immediate environment and the significant features of their photographs. The travel route, photographs and interview data were all geocoded. This produced a visual integrated narrative of the data that could be layered onto one another, allowing the researcher to present findings in a creative and impactful way (see Bergeron, et al., 2014, for examples).

Arts-based methods such as children's illustrations, maps and photographs have also been employed alongside go-along interviews (Asanin Dean, 2012; Bergeron, et al., 2014; Jones, et al., 2008). For example, Pawlowski and others (2014) explored the use of go-along interviews in examining barriers of recess physical activity for children. In this case, the researcher conducted go-along interviews while plotting locations of various places used, activities experienced, and children played with through an aerial photo of the schoolyard. This approach allowed the researcher to gain a better idea of where activities were taking place visually.

Indeed, go-along interviews are a method that brings together other commonly used data collection techniques in qualitative research, as well as combining other quantitative, geospatial and participatory approaches to meet the needs of the research objectives. Given the complexity of many areas of research in health geography, this novel method may prove especially helpful.

Using go-along interviews in health geography research

As a data collection tool, go-along interviews have the potential to provide insight into several emerging research areas and theories in health geography. While the contributions of this method cross-cut all three of these recent, and overlapping, trends in the subdiscipline, they will be discussed individually here.

New mobilities paradigm

The wider social sciences field has witnessed a growing interest in the movement of people and their bodies across micro and macro scales, as well as the movement of things and ideas including the related transportation systems, economic conditions, migration and citizenship processes, and advances in technology and communication (Cresswell, 2006; Sheller and Urry, 2006; Blunt, 2007). The impetus to investigate how movement influences the social life and experiences of place and vice versa has been the focus of the 'mobilities turn'. More specifically, Sheller and Urry (2006) suggest that this paradigm shift emphasizes the complexity of the people and place relationship stating that places 'are not so much fixed as implicated within the complex networks by which hosts, guests, buildings, objects and machines are contingently brought together to produce certain performances in certain places at certain times' (p. 214).

Within health geography, several researchers have called upon focusing on the interplay between well-being and mobility (Andrews, et al., 2012; Gatrell, 2013). Using the example of walking, Gattrell (2013) emphasizes the therapeutic benefits of mobility, including improved physical and mental health, social interaction, and positive affect associated with place. Similarly, Andrews and colleagues (2012) offer a critical look at the abundant walkability research and note that there is a significant absence of research that looks at the embodiment, mobility and movement activities of diverse individuals inhabiting or passing through these walkable places. They advocate for the engagement of health geographers in mobile activities research in order to untangle the relationship between place and people, such as (1) human made physical features and built environment, including land-use and urban design, (2) environmental factors including climatic and topographic conditions, (3) sociocultural features such as perceived crime or predominance of organized movement activities (e.g., running clubs), (4) sociopolitical conditions such as municipal active transportation priorities, and (5) cognitive and perceptual concerns including mobility for therapeutic purposes or public performance.

There are a number of ways in which go-along interviews can be used as a data collection tool to explore various facets of mobility and mobile activities. Most obvious is that go-along interviews are themselves a mobile method (Hein, et al., 2008) that allow for both the observation of others' mobility and personal experience of movement through a certain space and time. Through the interview, participant observation and fieldwork components of go-alongs interviews, all five areas of mobile activities research proposed by Andrews and colleagues (2012) can be examined. For instance, go-alongs can shed light onto the existence of and engagement with those aspects of the built environment intended to encourage safe mobility (e.g., parks, use of cross-walks) and the potential influence of environmental conditions (e.g., weather, traffic, perceptions of safety) on mobile activities (Asanin Dean, 2012).

The interview and observational components of go-alongs can provide insight into sociocultural contexts and even certain sociopolitical factors of place that influence mobility. In go-alongs with parents, I learned that they would take a particular route along a major road artery only if they were not walking with their children; in which case they would pick a 'safer' route (ongoing). Similarly, adolescent go-along participants often had rules from teachers about where in the community they could travel during lunch and recess breaks, which resulted in both increased and decreased mobility over the lunch break depending on the participant's inclination to follow or challenge these rules. Go-alongs with key informants (planners, councilors, public health specialists) could shed further light onto the larger scale sociopolitical factors influencing mobility. The potential for group go-alongs similar to those conducted by METRAC and Jane's Walk, can be utilized by researchers, practitioners (e.g., planners) and residents and act as a direct conduit to local level policy change.

Finally, go-alongs can certainly contribute to the cognitive and perceptual experiences of mobile activities through asking questions of participants about the drivers of their own behaviors, as well as providing opportunities for reflections.

As a go-along interviewer I engaged with local environments in ways I had not previously done in an attempt to experience place in similar ways as my participants. The most memorable experience was curb-walking, which one participant described as 'fun' and others remarked as 'something I always do'. From my vantage point, it seemed to be both an activity in balance and agility, as well as an activity of defiance (using neither the dedicated space for people or cars) and risk (one misstep away from falling into traffic). This was how these adolescents moved through their local environment and was just as important as the content of their interview answers, the total number of parks we passed and their perceptions about how safe the community was. I later tried curb-walking in my home neighborhood and was honked at almost immediately and subsequently asked if I needed help. This speaks to the sociocultural norms influencing mobility as my neighborhood seemed less accepting of this type of movement, and the reactions of others startled me, made me uncomfortable and slightly anxious about being known in the community as a deviant and risky walker – so I stopped. However, now as a seasonal cyclist who also gets honked at in my car-dependent suburban neighborhood, I am far less likely to stop this form of movement despite the horns and comments from drivers. The increasing social acceptance of cycling, the new cycling infrastructure in my community and my own priorities for being an active commuter when possible, I perceive as being in my favor, and though at times I would prefer to (illegally) cycle on the sidewalk on busy streets, I stay on the dedicated cycling lanes as both an act of compliance and support for the new active transportation priorities in my city but also of protest against the suburban status quo of car dependence and car-centric mind set of many of my neighbors.

Such acts of rebellion and protest are important to acknowledge as they do partially dictate how individuals move through space and engage with their local environments. Those interested in the planning and design of these environments could benefit from the insights that go-along interviews provide in order to create safer and more user-friendly cities.

Non-representational theory

Recently there has been a discussion about the potential role of non-representational theory in health geography (Andrews, et al., 2014; Hanlon, 2014; Kearns, 2014; Andrews, 2015) (see also Andrews and Drass in this volume). A non-representational approach, initially advocated by Nigel Thrift two decades ago (Thrift, 1996), is based on the premise that the widely used social constructivist perspective in social research is limited in its ability to capture, isolate, or understand the happenings of everyday life before they are given meaning and are represented (e.g., in speech). The emphasis of human geographers has been representational and has overlooked the importance of embodiment, performance and precognitive states (e.g., senses, expression) that may be more influential to our experience of the world rather than how we interpret it (Thrift, 1996; 2008; Cadman, 2009). A non-representational focus further examines how human and non-human subjects relate and co-evolve to create our experiences of the world (Thrift, 2008).

Andrews and colleagues (2014) suggest that one way to incorporate non-representational theory into the subdiscipline is through a focus on geographies of well-being and affect. Using ethnographic approaches in their own lives, they acknowledge that energetic, sensory then emotive experiences, felt in their immediate surroundings, contribute to an experience of well-being that emerges from – rather than an outcome of – an 'affective environment'. They state,

> The current study forwards quite different yet complementary understandings of wellbeing connected to it being – rather than a state of life and/or something found in prescribed situations – anchored in a common familiar feeling state. . . . In other words, wellbeing might initially emerge as an affective environment; the environmental action, then feeling of that action, prior to meaning. (Andrews, et al., 2014, p. 219)

In the analysis of their own everyday experiences, the authors emphasize sensory receptions of their environment (e.g., sounds of balls being thrown; vivid images at a water park; the steady movement of a quiet, cruising car; the heat, stuffiness and monotony of a waiting room ended by your name being called) and connect those experiences with their subsequent emotional state (e.g., excited, joyful, calm, relieved), which in turn impacts their well-being (Andrews, et al., 2014). While there has been some debate about the ability of any researcher to truly capture pre-cognitive experiences or name an action-induced emotion before it is given meaning (Hanlon, 2014; Kearns, 2014), there certainly is benefit to considering the role of an affective environment on well-being, and go-along interviews offer a potential mechanism to do so. Specifically, the interviewer can ask participants about their feeling states at various places in their neighborhood, or they can observe body language at given points during the interview. Go-alongs may prove insightful for researchers' own feeling states as they navigate an environment through the lens of someone else. One of my field notes offers a mix of these options:

> We were walking down a main road and there were two people who looked like they were in an argument on the corner. I thought I could hear that it was drug-related I wanted to cross the street and avoid the situation. I was feeling nervous about this situation escalating, as we got closer. What if the light was red at the corner and we had to stand there beside the arguing couple? I was starting to sweat now and I worried about how awkward it might be to have an audience during such a discussion. My participant was talking about health class at school and seemed unaware of what was up ahead. As we got closer I was eager to cross the road and was about to make a generic comment to the participant 'maybe we should go to the other side?' before I thought better of it. I wanted to experience this environment just as she did.

> This was her space. Her environment. My nerves should not influence the outcome of this interview. As we got closer to this couple, I looked over to the participant and there was not an ounce of fear or discomfort on her face. She looked more defiant and protective than fearful or tentative. And I was reminded that my job was to observe and participate; not direct her to walk her neighborhood in a certain way, or be fearful of strangers who had every right to inhabit space that same way we did. She stopped talking but walked a bit faster with her head up and I had struggled to keep up. As we approached, the couple didn't even notice us and the participant had intended to turn the corner anyway so we were not waiting for a light to change after all. We were of course fine and as we walked away from the corner, the participants said 'There are lots of pimps and hoes who work here.' Then she continued on her way telling me about friends at school. (Field note 14).

This research was not focused on feeling states, and I paid very little attention to them in my field notes, however I do remember thinking that as I looked at the participant, she showed a defiant and proud look typical of adolescents. While her performance may solely have been for my benefit – a point which underscores the complexity of interpreting behavior during go-alongs – my experience in that moment was that she did not seem phased by the arguing couple. I on the other hand was nervous and willing to change my activity pattern to avoid what was already an uncomfortable situation for me. I have no doubt that someone could have observed my body language and facial expression and known that I was not as confident and unphased by the situation as my walking partner. If we tune in enough, we can look at the participants, their bodies and watch their behavior to gain insight into how they are experiencing an affective environment. We can inquire about their emotions and feelings at that moment or even at the end of the interview when we can ask about emotions attached to events or places featured during the walk.

There are other suggestions for how to incorporate non-representational theory into the subdiscipline including through a focus on the investigation of materiality (Hanlon, 2014), and co-evolution (Hanlon, 2014; Kearns, 2014) (which will be covered more in depth below), and go-along interviews offer many potential insights into those processes.

Relational theory

A third and overlapping area of inquiry of interest to health geographers interested in go-along methodology is relational theory. Cummins and colleagues (2007) have called for the adoption of relational theory – places as dynamic and fluid network nodes with multiple social and cultural meanings rather than fixed boundaries with static or neutral sociocultural interpretations – by geographers researching health inequalities. These authors and others (Cattell, et al., 2008; Davidson, et al., 2008; Andrews, et al., 2012) suggest that improved understandings of the health and place relationship is dependent on dissolving the context (e.g., area level characteristics such as community resources, pollution, social capital) – composition (e.g., individual characteristics such as gender, age, income) and binary, and instead focus on the interaction between people and place. However, empirically investigating the interrelationships between context and composition is challenging (Cummins, et al., 2007).

The incorporation of GPS technologies to track people's activity space such as those studies noted above, provide insight into how use of physical space varies across populations, over time and by location (e.g., level of urbanization, presence of amenities), but these technologies alone lack the necessary insight into experiences of place in a way to truly understand relationality. I echo Carpiano's (2009) sentiments that the go-along interview 'provides a unique way for the researcher to not only observe people's neighborhood environments, but to also study people's perceptions, processing, and navigation of their environments' (p. 264), that is, how they engage with and experience their neighborhood.

In the adolescent body weight study, several participants were unable to recall street names or identify the name or purpose of buildings in their neighborhoods, but were nevertheless able to identify the houses where their friends or family lived, or where they used to live. Their preferred walking routes were determined not necessarily by distance or aesthetics, but by their experiences or perceptions of those spaces as one participant suggests:

> We don't really visit this street down here anymore [points to side street] because there is a guy who likes little children, and does bad things to them . . . so I don't go down his street because I am actually afraid something might happen, so if I were to go see my friend who lives over there [points to house near said street], I would walk on the other side. One time [the friend] had asked me to go over to his house and I was like, 'Dude, no way. I am staying over here.' (15 year old female)

For adolescent participants in the study, it was the socio-relational aspects of their local environment more than the physical attributes that influenced how, when and why they used local space. The ability to witness and participate in these experiences as a researcher was made possible through the go-along interview method.

Go-along interviews challenge the deterministic view of the environment in shaping human behavior, and I have found them especially insightful when working on projects with interdisciplinary teams where theories of rational choice (Levinson and Zhu, 2013) and architectural (neoenvironmental) determinism (Ewing, 2002) are commonplace. The findings from go-along interviews have contributed to increased understandings among team members of the importance of examining how people interpret, use and justify their engagement with local environments. As a social scientist and health geographer, I firmly believe the overall project benefits from a more relational approach to examining the role of built environments on human behavior, and this knowledge also often shifts perceptions of practitioners and other team members about the context and composition interaction.

Finally, the power of go-along interviews sheds light onto relational theory that may be enhanced when combining it with other methods of data collection such as GPS tracking, visual mapping or subsequent interviews to better understand the co-evolution and interaction between context and composition.

Conclusion

In summary, go-along interviews have much promise as a novel method of data collection in health geography. This method brings together the strong methodological roots of ethnography and the explanatory power and depth of individual interviews and has a lot to offer the future of health geography as the subdiscipline moves forward in exploring new paradigms and theories to better understand the relationship between health and place.

References

Andrews, G. J., 2015. The lively challenges and opportunities of non-representational theory: a reply to Hanlon and Kearns. *Social Science and Medicine*, 128, pp. 338–341. doi:10.1016/j.socscimed.2014.09.004.

Andrews, G. J., Chen, S., and Myers, S., 2014. The 'taking place' of health and wellbeing: Towards non-representational theory. *Social Science and Medicine*, 108, pp. 210–222. doi:10.1016/j.socscimed.2014.02.037.

Andrews, G. J., Hall, E., Evans, B., and Colls, R., 2012. Moving beyond walkability: on the potential of health geography. *Social Science and Medicine*, 75, pp. 1925–1932. doi:10.1016/j.socscimed.2012.08.013.

Asanin Dean, J., 2012. *Local environmental determinants of adolescent body weight in low-socioeconomic status neighbourhoods in Ontario, Canada*. PhD dissertation, McMaster University, Hamilton, Ontario.

Bergeron, J., Paquette, S., and Poullaouec-Gonidec, P., 2014. Uncovering landscape values and micro-geographies of meanings with the go-along method. *Landscape and Urban Planning*, 122, pp. 108–121. doi:10.1016/j.landurbplan.2013.11.009.

Blunt, A., 2007. Cultural geographies of migration: mobility, transnationality and diaspora. *Progress in Human Geography*, 31, pp. 684–694.

Bors, P. A., Brownson, R. C., and Brennan, L. K., 2012. Assessment for active living: harnessing the power of data-driven planning and action. *American Journal of Preventive Medicine*, 43, pp. S300–S308. doi:10.1016/j.amepre.2012.06.023.

Cadman, L. 2009. Nonrepresentational theory/nonrepresentational geographies. *International Encyclopedia of Human Geography*, 7, pp. 456–63.

Carpiano, R. M., 2009. Come take a walk with me: the 'go-along' interview as a novel method for studying the implications of place for health and well-being. *Health and Place*, 15, pp. 263–272. doi:10.1016/j.healthplace.2008.05.003.

Cattell, V., Dines, N., Gesler, W., and Curtis, S., 2008. Mingling, observing, and lingering: everyday public spaces and their implications for well-being and social relations. *Health and place*, 14(3), pp. 544–561.

Chow, B., 2011. *'Everybody else got to have this cookie': the effects of food allergen labels on the well-being of canadians*. Master's thesis, McMaster University, Hamilton, Ontario.

Cresswell, T., 2006. *On the move: mobility in the modern western world.* New York: Taylor and Francis.

Cummins, S., Curtis, S., Diez-Roux, A. V., and Macintyre, S., 2007. Understanding and representing 'place' in health research: a relational approach. *Social Science and Medicine*, 65, pp. 1825–1838. doi:10.1016/j.socscimed.2007.05.036.

Dean, J. A., and Elliott, S. J., 2012. Prioritizing obesity in the city. *Journal of Urban Health*, 89, pp. 196–213. doi:10.1007/s11524-011-9620-3.

Davidson, R., Mitchell, R., and Hunt, K., 2008. Location, location, location: the role of experience of disadvantage in lay perceptions of area inequalities in health. *Health and Place*, 14(2), pp. 167–181.

Dyer, S., and Demeritt, D. 2009. Un-ethical review? Why it is wrong to apply the medical model of research governance to human geography. *Progress in Human Geography*, 33(1), pp. 46–64. doi:10.1177/0309132508090475.

Eisenberg, M. E., Garcia, C. M., Frerich, E. A., Lechner, K. E., and Lust, K. A., 2012. Through the eyes of the student: what college students look for, find, and think about sexual health resources on campus. *Sexuality Research and Social Policy*, 9, pp. 306–316. doi:10.1007/s13178-012-0087-0.

Evans, J., and Jones, P., 2011. The walking interview: methodology, mobility and place. *Applied Geography*, 31, pp. 849–858. doi:10.1016/j.apgeog.2010.09.005.

Farmer, T., Robinson, K., Elliot, S., and Eyles, J., 2006. Developing and implementing a triangulation protocol for qualitative health research. *Qualitative Health Research*, 16, pp. 377–394.

Garcia, C. M., Eisenberg, M. E., Frerich, E. A., Lechner, K. E., and Lust, K., 2012. Conducting go-along interviews to understand context and promote health. *Qualitative Health Research*, 22, pp. 1395–1403. doi:10.1177/1049732312452936.

Gardner, P. J., 2011. Natural neighbourhood networks – Important social networks in the lives of older adults aging in place. *Journal of Aging Studies*, 25, pp. 263–271. doi:10.1016/j.jaging.2011.03.007.

Gatrell, A. C., 2013. Therapeutic mobilities: walking and 'steps' to wellbeing and health. *Health and Place*, 22, pp. 98–106. doi:10.1016/j.healthplace.2013.04.002.

Haggerty, K. D., 2004. Ethics creep: governing social science research in the name of ethics. *Qualitative Sociology*, 27(4), pp. 391–414. doi:10.1023/B:QUAS.0000049239.15922.a3.

Hamilton 2015. Neighbourhood action strategy. Available at: http://preview.hamilton.ca/city-initiatives/strategies-actions/neighbourhood-action-strategy. (Accessed May 14, 2015.)

Hanlon, N., 2014. Doing health geography with feeling. *Social Science and Medicine*, 115, pp. 144–146. doi:10.1016/j.socscimed.2014.05.039.

Hein, J. R., Evans, J., and Jones, P., 2008. Mobile methodologies: theory, technology and practice. *Geography Compass*, 2, pp. 1266–1285. doi:10.1111/j.1749-8198.2008.00139.x.

Jane's Walk. 2015. Jane's Walk. http://janeswalk.org. (Accessed on May 14, 2015.)

Jones, P., Evans, J., Gibbs, H., and Hein, J., 2008. Exploring space and place with walking interviews. *Journal of Research Practice*, 4, pp. 1–9.

Juritzen, T. I., Grimen, H., and Heggen, K. 2011. Protecting vulnerable research participants: a Foucault-inspired analysis of ethics committees. *Nursing Ethics*, *18*(5), pp. 640–650. doi:10.1177/0969733011403807.

Kearns, R. A., 2014. The health in 'life's infinite doings': A response to Andrews et al. *Social Science and Medicine*, 115, pp. 147–149. doi:10.1016/j.socscimed.2014.05.040.

Kelly, C. E., Tight, M. R., Hodgson, F. C., Page, M. W., 2011. A comparison of three methods for assessing the walkability of the pedestrian environment. *Journal of Transport Geography*, 19, pp. 1500–1508. doi:10.1016/j.jtrangeo.2010.08.001.

Kusenbach, M., 2003. Street phenomenology: the go-along as ethnographic research tool. *Ethnography*, 4, pp. 455–485. doi:10.1177/146613810343007.

Levinson, D., and Zhu, S., 2013. A portfolio theory of route choice. *Transportation Research Part C: Emerging Technologies*, 35, pp. 232–243.

Metro Action Committee on Public Violence against Women and Children, 1989. *Women's safety audit kit guidebook*. Toronto, Ontario: METRAC.

Paquet, C., Daniel, M., Kestens, T., Léger, K., and Gauvin, L., 2008. Field validation of listings of food stores and commercial physical activity establishments from secondary data. *International Journal of Behavioural Nutrition and Physical Activity*, 5, p. 58.

Patton, M. Q., 2002. *Qualitative research and evaluation methods*. 3rd ed. Thousand Oaks, CA: Sage.

Pawlowski, C., Tjørnhøj-Thomsen, T., Schipperijn, J., and Troelsen, J., 2014. Barriers for recess physical activity: a gender specific qualitative focus group exploration. *BMC Public Health*, 14, p. 639. doi:10.1186/1471-2458-14-639.

Sampson, R. J., and Raudenbush, S. W., 1999. Systematic social observation of public spaces: a new look at disorder in urban neighborhoods. *American Journal of Sociology*, 105, pp. 603–651. doi:10.1086/210356.

Sharkey, J. R., and Horel, S., 2008. Neighborhood socioeconomic deprivation and minority composition are associated with better potential spatial access to the ground-truthed food environment in a large rural area, *Journal of Nutrition*, 138, pp. 620–627.

Sheller, M., and Urry, J., 2006. The new mobilities paradigm. *Environment and Planning A*, 38 pp. 207–226.

Thrift, N., 1996. *Spatial formations*. Vol. 42. London: Sage.

Thrift, N., 2008. *Non-representational theory: Space, politics, affect*. New York: Routledge.

Part III

Representation through visual media

8 What can participant-generated drawing add to health geography's qualitative palette?

Stephanie E. Coen

Reflections on praxis

Deploying participant-generated drawing as a research method presented me with two new challenges. The first is rather personal. Although I am convinced that drawing is a valuable method and stand by what I have written in this chapter, I wrestled with the internal dilemma that I am far from a drawer myself. Dabbling in drawing as a method has helped me to realize just how very much I think *through* writing. I literally formulate my ideas by putting words down on the page and moving them around until concepts and arguments take shape. From the outset I was aware that drawing might not be the best method for me if the roles were reversed – indeed part of my rationale for piloting this method was to engage with this dissonance. Two realizations have helped me to contend with this discrepancy. For one, my findings make clear that there are a number of methodologically significant reasons to use drawing to explore men's and women's experiences in gym environments and in health geography research more widely. No single qualitative method is a perfect fit for everyone and it could be equally problematic to avoid drawing simply because of my own personal communicative preferences if it is a productive method for the research question. Next, the overall experience of my participants with drawing was positive – even for the one who preferred *not* to draw! This participant was able to embrace the activity yet still honor her preference for words (this could have been me!).

My second challenge related to analysis. Despite exploring numerous ways to interpret visual data using critical approaches, I could not find a technique that made me feel in any way equipped to interpret someone else's drawing. I resolved this by taking an auteur theory approach consistent with the feminist orientation of my research that privileges participant perspectives and acknowledges their expertise (Dyck, 1999). I still question, though, what – if anything – could be gained from engaging more directly with the visual component in addition to participants' own interpretations. If 'a picture is worth a thousand words', what happens when participants re-represent their images in words and researchers then parse these texts into themes? These challenges are not inherently problematic, simply illustrative of some of the opportunities for critical reflexivity implicated in drawing.

Introduction

A growing body of qualitative health research demonstrates that drawing as a research method can yield important insights into a diversity of adults' health-related experiences and perceptions from HIV testing (Mays, et al., 2011) to living with spinal cord injury (Cross, Kabel and Lysack, 2006). In this chapter, I explore what drawing can add to health geography's qualitative palette. Participatory mapping techniques familiar to health geography may involve drawing, but such applications are more often limited to representations of cartographic or spatial knowledge rather than more flexible depictions of the experiential dimensions of health. The form of drawing that I consider here is participant-generated, meaning that it involves asking participants to draw in response to a question (or questions) and then to describe or interpret the resulting image (or images). This mode of drawing is typically engaged in conjunction with more conventional social research methods, such as interviews (Guillemin, 2004). Unless stated otherwise, drawing throughout this chapter refers to the participant-generated variety, as opposed to researcher-generated or use of other existing drawings. In the first part of this chapter, I look to literature outside of geography to illustrate the possibilities that drawing presents to advance some of the methodological commitments in qualitative health geography to foreground first-hand experience and take seriously everyday contexts relevant for health (cf. Kearns, 1993; Dyck, 1999). I then report on a pilot study in which I sought to empirically assess the suitability of drawing as a method for my research on women's and men's experiences in gym environments and to evaluate the benefits and constraints of drawing from the perspectives of research participants. Overall, I argue that there is a strong alignment between the kinds of knowledge that drawing can help to produce and the types of topics that many geographies of health seek to understand. I suggest that the methodological utility of drawing may lie not only in advancing *what* we know, but in *how* we do health geography research. As such, this method may offer added benefits for a range of health geographers committed to privileging first-hand accounts of health-related experiences, in particular those working from critical or feminist perspectives. At the same time, I underline that despite its apparent novelty, drawing should not be implemented uncritically.

I first set out to study gyms in response to critiques levied by health geographers that geographical research on physical activity had been largely asocial, for the most part subscribing to ecological approaches emphasizing the built environment (Andrews, et al., 2012; Colls and Evans, 2013), often to the neglect of people's experiences within the distinct sites and facilities where individuals take up physical activities (Andrews, et al., 2012; Blacksher and Lovasi, 2012). Gyms are one such increasingly common place, yet they can reinforce and perpetuate problematic gender stereotypes about both men and women. Research has shown that gyms can be divisive along gender lines, with weight rooms perceived as masculine and 'cardio' areas as feminine (Johnston, 1998; Dworkin, 2003). From a geographical perspective, however, we know little about the socio-spatial processes that work to exclude or include people in various workout spaces. By focusing on these microdynamics, I hoped to identify points for intervention to make gym environments more supportive of the participation of a diversity of users.

My methodology is informed by the stream of health geography that takes a social conception of health, is concerned with the dynamic relationships between place and health, and holds the experiential elements of these relationships to be important (Kearns, 1993; Dyck, 1999). Methods stemming from this approach allow for 'grounding of research in the everyday locales where health care practices and health behavior are "played out"' (Dyck, 1999, p. 246) and are open to unsettling taken-for-granted social categories (Dyck, 1999). Methodologically though, I grappled with how to get at the experiential aspects of this arguably mundane, yet significant, physical activity context. Gender too is often taken-for-granted in everyday life as 'natural' and unquestioned (West and Zimmerman, 1987). Given the seeming ordinariness of both the gym and gender, I was concerned that participants might deem aspects of their gym experience too obvious or inherent to mention in an interview alone.

In addition, I conceptualized the gym as contentious site for the practice of 'good' health behavior within in a wider context of increasing individualization of personal health responsibility (Crawford, 1980; Brown and Duncan, 2002), or what Crawford (1980) calls 'healthism.' I saw this health imperative, with all its moral trimmings, not only as an important backdrop for what happens inside the gym, but also for what people might tell me about it. Finally, I was cognizant that my own positionality (Rose, 1997), particularly as a long-time exerciser, one-time competitive powerlifter and group indoor cycling instructor, shaped the questions I was posing – even the fact that I was doing this research at all. How could I encourage my participants to re-examine the seemingly obvious? What could I do to assuage a sense of obligation that participants might feel to demonstrate 'good' health behavior to me? How could I push myself outside my own gym frame of reference to avoid foreclosing potentially important questions?

In seeking a solution, I came across a small but compelling body of health research using drawing as an adjunct to augment more familiar qualitative research methods. Particularly intriguing to me was how a few studies enlisted drawing as a way to investigate topics where dominant social scripts might be particularly at play in interview settings. For example, Mays and co-authors (2011) included drawing in their research on reasons for declining HIV testing because they suspected that participants might feel pressure to provide socially acceptable answers in standalone interviews.[1] Similarly, in their study on recovery from addiction, Shinebourne and Smith (2011) turned to drawing out of anticipation

1 While various techniques to avoid what is referred to as 'social desirability bias' have been developed for more quantitative social survey research (Krumpal, 2013), to the best of my knowledge such techniques have not received a great deal of attention in qualitative interviewing, perhaps because this is less problematic when we consider research knowledge to be always situated or partial rather than objective or value-free (Haraway, 1988; Rose, 1997).

that interview dynamics might compel participants to replicate and adhere to the language of treatment programs. These questions and possibilities inspired my foray into drawing.

Drawing as a research method

Before discussing the results of my pilot study, I first present a primer on drawing as a research method. I locate drawing within the wider family of visual methodologies, differentiate between drawing techniques, identify approaches to analysis, discuss several methodological benefits, and point to some potential pitfalls.

Locating participant-generated drawing in visual methodologies

Visual methodologies are united in the epistemological assumption that the visual constitutes a particular sensory way of knowing and representing the world (Rose, 2001). Proponents contend that visual methods provide entry points into an array of research questions that may not be accessible via other (non-visual) senses or modes of expression (Gauntlett and Holzwarth, 2006; Banks, 2008). Yet, there are a diversity of approaches to contending with visual content and culture (Rose, 2001). It is important to distinguish between those that utilize researcher-created or existing imagery versus approaches that employ participant-generated visuals because they entail very different research processes (Guillemin and Drew, 2010). The latter domain, where the form of drawing that I consider fits in, engages participants in the research process as active producers, and oftentimes interpreters, of visual products (Guillemin and Drew, 2010). In the wider field of visual methods, the participant-generated suite of visual techniques has largely been dominated by photographic approaches, such as photo elicitation and photovoice (cf. Harrison, 2002; Azzarito, 2010). Photographic methods, however, comprise a particular visual genre with its own methodological and epistemological issues (such as use of technology and representation of 'real-world' scenes) that may not be appropriate for all health-related topics, yet visual approaches could still be beneficial. This may be the case if a topic is of a highly sensitive nature or emotionally difficult to speak about (Guillemin and Westall, 2008; Scott, 2009; Guillemin and Drew, 2010) or where for ethical, practical, or privacy reasons photography may be prohibited, such as in gyms.

Drawing has gained increasing attention in health research as a participant-generated visual method with distinct methodological attributes (cf. Guillemin, 2004; Guillemin and Westall, 2008; Guillemin and Drew, 2010). What sets drawing apart, according to Guillemin (2004) who has extensively employed drawing in her research on women's heart disease, postnatal depression and menopause, is its dual role as both verb and noun; drawing is at once *process* (the act of drawing) and *product* (the drawing itself) (p. 274). For Guillemin, this is no linguistic coincidence, but helps to frame drawing as a meaning-making activity that generates visual representations of experiences as understood in particular places and times. As process, the act of creating something from scratch on a blank sheet of paper necessitates a degree of thoughtful engagement on the part of participants that may uniquely stimulate knowledge construction (Guillemin, 2004; Shinebourne and Smith, 2011). In addition, unlike other visuals such as photographs, drawing is not necessarily tied – or perceived to be tied – to the depiction of 'real' images. Drawings may be abstract, concrete or anything in between. As such, this medium may engage participants in a qualitatively different way than capturing images via other means and offers a categorically different platform for research participant expression (Harrison, 2002). As product, drawing provides a stepping-stone to foster discussion with research participants about the topic at hand (Guillemin and Drew, 2010; Shinebourne and Smith, 2011). Viewing the resulting image may spur participants to reflect further on their experiences (Guillemin, 2004; Morgan, et al., 2009).

Locating drawing within the broader family of visual methods also requires attending to the relationships between the visual and other senses and systems of communication. Drawing, like all visual methods, does not exist in a visual vacuum, but is always interrelated with other forms of knowledge. Geographer Gillian Rose (2001), in her framework for a critical visual methodology, notes that even though the visual might not be reducible to language alone, it exists relationally with other texts. This dynamic is particularly pertinent to drawing as method as it is most often used in conjunction with other (verbal or written) social research methods (Guillemin, 2004). The interpretation of drawing data also inevitably circles back to the linguistic realm regardless of whether images are interpreted by participants or researchers or both (Guillemin, 2004; Cross, et al., 2006). While drawing is categorized as a visual research method, its application requires engagement with more traditional non-visual data collection and analysis techniques, such as interviews and thematic coding, and the well-established methodological considerations that go along with them.

Operationalizing and analyzing participant-generated drawing

There are several variations to consider in determining how to operationalize drawing into a research methodology. First, the number of questions or drawing prompts can vary. Some studies employ a single open-ended question to solicit a single image. Guillemin (2004), for example, uses single questions, asking women in one study to draw 'how they understand their menopause' (p. 276) and women in separate study to draw how they 'visualized their heart disease' (p. 276). Other researchers use multiple questions to generate multiple drawings. In their work on women's experiences with chronic vaginal thrush, Morgan and colleagues (2009) asked women to draw 'how chronic vaginal thrush has made you feel in your day to day life' (p. 131), followed by additional requests for separate images about their experiences with biomedical and complementary and alternative medicine treatments. More than one drawing also presents the possibility to pose comparative questions. For instance, Scott (2009), in her study of AIDS meanings asked HIV-positive women to draw '"a picture she had in her mind" of HIV and how it was in her body' (p. 456) followed by an image of what AIDS looks like. She then engaged participants in a discussion of the difference between the two. Second, the researcher needs to choose the placement of the drawing exercise within an interview. Guillemin (2004) argues that placing the drawing activity at the conclusion of an interview is crucial to establish the level of rapport needed to facilitate drawing. Indeed, Cross and co-authors (2006), when studying experiences of living with spinal cord injury went so far as to implement drawing at the end of a second round of interviews with participants, asking them to 'draw yourself' and then 'draw how you see your spinal cord injury in your mind' (p. 185). In contrast, Shinebourne and Smith (2011) opened with drawing, inviting participants to first draw for ten minutes about how they presently see their recovery from addiction before commencing an interview. Third, drawing techniques can be distinguished by where the visual activity and the verbal interpretation meet. While most applications of drawing ask participants to describe and interpret the finished drawing product (e.g., Cross, et al., 2006), some researchers pose questions to participants during the drawing process (e.g., Scott, 2009). Relatedly, a key consideration in operationalizing drawing is whether the researcher is present while the participant is producing the drawing. Guillemin and Drew (2010) suggest that researchers should be critically reflective about how their presence (or absence) may affect drawing production. If the researcher becomes audience to the image production, do participants feel watched or compelled to perform drawing in certain ways? A fourth consideration is the duration of the drawing exercise. While the time spent drawing is most often left up to participants, an exception is Shinebourne and Smith (2011), who ask participants to draw for a set period at the onset of the research encounter. Lastly, researchers are faced with nearly endless options about the types of drawing materials to be made available for participants. This choice is more or less important depending upon whether the analytic strategy is strictly textual or includes a visual analysis of the drawing itself by, for example, extending to consider color and composition.

In terms of practically implementing drawing, as with other qualitative methods, drawing requires significant rapport between participant and researcher. As mentioned, drawing at the end of an interview may be one way to ensure sufficient rapport is in place (Guillemin, 2004; Guillemin and Westall, 2008; Morgan, et al., 2009). Participants may also initially feel tentative about drawing, and either hesitate or decline. To address this, Guillemin (2004) suggests ensuring participants are well-informed beforehand, being patient and offering encouragement, and reiterating that artistic skills are irrelevant. Defusing any perceived artistic pressure is paramount to ensuring that participants are not overwhelmed by the request. Cases where participants ultimately refuse to draw should be subject to analytic scrutiny (Guillemin, 2004).

Analyzing drawing data largely relates to how two related questions are addressed in study design. First, what is the role of the participant's interpretation of their drawing and what is the role of the researcher's interpretation of the drawing? Some researchers argue that the drawer's interpretation of their own image should take primacy, and that the role of the researcher is instead to analyze and synthesize across interpretations within a given study (e.g., Gauntlett and Holzwarth, 2006; Guillemin and Drew, 2010). Holding the maker's intention as the most important element in the interpretation of images is a perspective that Rose (2001) classifies as 'auteur theory,' to which I subscribe in this chapter. Others allow for greater variation in the author-to-researcher interpretation ratio, especially if the researcher has undertaken a separate visual analysis of the images (e.g., Cross, et al., 2006). Second, does the analysis contend with textual data, visual data or both? Some researchers, for example, use drawings primarily as a tool for talk, incorporating participants' oral descriptions into the main interview narrative and analyzing them as part of the interview text (e.g., Morgan, et al., 2009; Mays, et al., 2011). This is what Harrison (2002) refers to as the visual being 'a technical means to an end; that is, the generation of verbal data for analysis' (p. 865). Others perform visual analyses of the drawing products, in addition to any textual analysis of participant interpretations (e.g., Scott, 2009; Shinebourne and Smith, 2011). How visual and verbal data are analyzed likely depends on whether one takes the visual as topic (i.e., substantive focus of study) or tool (i.e., an entry point to the topic) (Harrison, 2002).

Methodological benefits

Certainly different qualitative methods are needed to query various aspects of understanding and experience, so how can drawing specifically diversify our methods palette? Below, I outline three methodological mechanisms by which drawing can benefit qualitative research practice in health geography, before discussing some limitations of the method. These themes are not neatly collapsible into discrete categories – they are wrapped up together in the messy process that is drawing. I separate these here as a heuristic device to conceptualize the methodological contributions of drawing.

Comprising a medium for exploring emotions

Emotions are an important part of unpacking the experiential and contextual dimensions of health of interest to health geographers seeking to foreground human agency and participant perceptions (Davidson and Milligan, 2004; Thien and Del Casino, 2012). An often cited strength of drawing is that it provides a means to access emotional aspects of experience that may not emerge in the context of a traditional interview, particularly those that may be ineffable or not easily spoken (Nowicka-Sauer, 2007; Morgan, et al., 2009; Guillemin and Drew, 2010; Reason, 2010; Shinebourne and Smith, 2011). For example, Nowicka-Sauer (2007), in her US work on women's experiences of living with lupus noted that drawings prompted participants to think differently – in a non-medical way – about their experiences, by 'concentrating on "feeling" and "experiencing" the disease' (p. 1524). Drawing may bring to the surface sensitive material that prompts participants to reconsider aspects of their experience in potentially emotionally charged ways. In their Australian research on postnatal depression, Guillemin and Westall (2008) observed that drawing revealed women's feelings of isolation, despair and entrapment that many women found difficult to articulate in interviews. Likewise, Cross, Kabel, and Lysack (2006), in their US study of community integration of adults with spinal cord injury, found that in interviews participants described adjusting well to changes in their life circumstances with spinal cord injury, while through drawing, participants expressed some of the unresolved issues and struggles they faced. Drawing can thus be a more palatable platform, than direct talk, for broaching sensitive topics. There are, however, ethical concerns if drawing spurs participants to relive difficult and painful aspects of their experiences. Researchers should take measures to pre-empt and minimize any such risk (Morgan, et al., 2009; Guillemin and Drew, 2010).

Such emotional content impacts not only participants and researchers, but also other audiences who view the resulting drawings. Drawings can provide vivid evidence to incite action by, for example, portraying patient perspectives in ways that can be illuminating for practitioners and health policymakers (Cross, et al., 2006; Morgan, et al., 2009). In their editorial on the implications of arts-based research for healthcare professionals, Parsons and Boydell (2012) argue that such research products can convey lived experience in ways that can powerfully communicate messages to target audiences in practice and policy arenas. The capacity of drawing to be a potent translational tool for research findings may be especially beneficial for critical health geographers seeking to inform policy and practice changes through their research.

Supporting a participant-centered research process

Drawing as a method can help to support a participant-centered research process, and this aligns well with the stream of qualitative health geography that subscribes to subject-centered approaches (Dyck, 1999). An important aspect that sets drawing apart from many talk- and word-based techniques – and indeed some other visual methods – is the potential time involved in the drawing process, although this ultimately depends upon how drawing is administered. As it is most commonly employed, drawing disrupts the question-answer rhythm of research. By not demanding an instantaneous response, as may a face-to-face interview, drawing allows room for reflection about the phenomena at hand (Guillemin and Drew, 2010). This has implications for the character of the data derived, as argued by Gauntlett and Holzwarth (2006, p. 84): 'The data you end up with is the result of thoughtful reflection.' This room for reflection arguably puts the locus of knowledge production more squarely in the participants' court. Participants may take more conscious control over the knowledge production process and more ownership over the research time. In this way, drawing may work to alter the terms of participant engagement with research, particularly if an auteur theory approach is taken to recognize participants as experts about their products and invites their interpretive contributions. The finished drawing product can also serve a reflective function. Shinebourne and Smith (2011) explain that 'the activity of drawing results in an enduring object which can be used as a resource for remembering the past as it presents itself in the present' (p. 320). This opportunity for in situ reflection breaks with traditional research dynamics and offers the potential for a participant-driven process (Guillemin and Drew, 2010).

Making sense of experiences and representing them through drawing can also generate nuanced or heightened self-understandings, often in the emotional realm as discussed above (Guillemin, 2004; Cross, et al., 2006; Gauntlett and Holzwarth, 2006). For example, in Morgan and colleagues' (2009) Australian study of women's use of complementary and alternative medicine to manage chronic vaginal thrush, participants reported that drawing helped them to stand back and see how they feel. From this vantage point, participants were able to articulate how they experienced a sense of disconnection from their physical bodies to cope with daily pain and discomfort. The opportunity to open up about a sensitive topic via drawing may be cathartic in itself (Morgan, et al., 2006). A number of studies identify therapeutic benefits as a positive side effect of drawing (Morgan, et al., 2006; Nowicka-Sauer, 2007; Guillemin and Drew, 2010). Such informal therapeutic effects may be an unintended benefit that researchers should consider, but I would caution researchers are not necessarily therapists. There is an important difference between therapeutic effects as positive by-products of drawing versus using drawing as a form of therapy – the latter being the domain of trained therapists and clinicians. Overall, drawing can foster a more participant-centered research process by disrupting embedded power differentials (e.g., time allocation, expertise), positioning participants to make meaningful analytical contributions to research (e.g., as interpreters of their images), and deriving benefits from being involved (e.g., self-awareness, therapeutic benefits).

Reinforcing rigor

Drawing can be valuable in 'sighting' similarities and contradictions across visual and linguistic modes of expression (Guillemin, 2004; Guillemin and Westall, 2008). This is particularly pertinent in research designs employing a triangulation strategy. Scott's (2009) research, which combined drawing with interviews and free lists, concluded that drawings reinforced other data types and allowed for a deeper exploration of HIV/AIDS meanings than would not have been possible without them. Other empirical examples demonstrate that drawing can lend credibility to researcher interpretations of other data types (e.g., Morgan, et al., 2009) or surface ambiguities and contradictions among them (Guillemin, 2004). As one more opportunity to probe data further and interrogate seeming discrepancies among visual and other modes of expression and experience, drawing can be a useful check for rigor (cf. Baxter and Eyles, 1997). It could also be conceived as a form of participant validation (cf. Turner and Coen, 2008) in which participants can augment or edit their interview responses based on any realizations post-drawing (Guillemin, 2004). Cross and team (2006), who conducted two separate interviews with each participant, intentionally placed the drawing activity at the end of the second interview as a way to encourage participants to fill in any gaps not previously covered (see also Guillemin, 2004). One participant, for example, who had indicated acceptance of his spinal cord injury, only revealed challenges he experienced associated with loss of independence, privacy, and control based on his drawing. While using drawing to reinforce rigor has implications outside of the subdiscipline of health geography, it does present a potentially innovative opportunity to advance rigor within qualitative health geography practices.

Beyond being a consciousness-raising activity for participants, drawing can be likewise for researchers. Mannay (2010) argues that participant-generated visuals (albeit not using drawing specifically, but photo-elicitation, collage-making, and mapping) can be a productive way to 'make the familiar strange'. In her research on mothers' and daughters' experiences living in a marginal UK neighborhood – the same neighborhood where she resided – Mannay makes the case that participant-created visual products can render visible preconceptions on the part of the researcher, as well as shine a light on assumptions about shared points of reference or sameness among researchers and participants. Given her familiarity with the research setting, Mannay found that certain interview questions were awkward and seemed irrelevant to pose. To get around this, she used participant-generated visuals as 'an instrument for making the familiar strange and . . . a gateway to destinations that lay beyond my repertoire of preconceived understandings of place and space' (p. 96). Mannay echoes a broader methodological argument put forth by Banks (2008, n.p.) that visual methods 'can and should provoke the researcher into (re)considering taken-for-granted analytical categories.' Drawing can thus be an especially useful analytic tool in situations where the researcher is close to the research setting and topic (Mannay, 2010). Drawing in this sense is more than a visual method for data elicitation; it can be a tool for critical self-reflection on the part of the researcher to support rigorous qualitative analysis. In health geography, this could be one strategy to engage in the sometimes enigmatic practice of researcher reflexivity.

Potential pitfalls

Capitalizing on the methodological strengths of drawing requires cognizance of its potential pitfalls. Given that many qualitative health geographies subscribe to social constructivist, critical, or feminist epistemologies, it is especially important that health geographers recognize its positivist heritage and applications. One of the primary methodological concerns about drawing is that it can be (and has been) used as if it could represent something absolute – and ultimately decontextualized – about people's lives and experience. This is the case with historical uses of the draw and write technique (drawing and writing in response to a question or prompt) in health research with children, from which drawing with adults has its origins (Backett-Milburn and McKie, 1999). Some of the foremost problems with draw and write include insufficient consideration of how the research process shapes what is drawn and a tendency to quantify drawing data (such as counting or measuring compositional elements), thereby applying a qualitative method toward quantitative ends (Backett-Milburn and McKie, 1999). Indeed, there is a recent vein of drawing in medical research with adults that exemplifies these tensions, using participant drawings to make correlations between the compositional properties of drawings and characteristics of participants. In studies on heart failure (Reynolds, et al., 2007) and myocardial infarction (Broadbent, et al., 2004), patients were asked to draw their hearts before and after their health incident. Drawings were then subjected to computer-based analysis which calculated the *size* of the hearts drawn and analyzed this measurement in conjunction with other survey data. These pictorial hearts were taken as indicators of the actual health status of individuals – an approach counter to the non-positivist epistemological underpinnings of many qualitative health geographies. Recent research employing drawing with children and adolescents in health geography (see Fenton, et al., 2011) has done so critically, but these methodological cautions still require carrying forward into health geography research with adults.

Relatedly, there is the danger that drawing can be undertaken as a projective technique, essentially using deception and lack of information to elicit data from the seeming 'depths' of participants' consciousness (Reason, 2010). Not only do such applications pose an ethical issue of lack of fully informed consent, but 'the claim that the methodology reveals truths below the surface of the participants' consciousness' ultimately positions participants as passive subjects (Reason, 2010, p. 394). Drawing in this way is disabling, as it implies a lack of agency on the part of participants (Reason, 2010). One way to counter this potential pitfall, according to Gauntlett and Holzwarth (2006, p. 82), is to situate visual methods within methodological frameworks that are 'optimistic and trusting about people's ability to generate interesting theories and observations themselves'. Adopting an auteur theory approach is one way to realign the balance of expertise and agency toward the participant (Rose, 2001; Reason, 2010). Soliciting participants' interpretations of their drawings also ensures that drawing data is not devoid of context. Indeed, for this reason, many researchers argue that drawing *must* be accompanied by participants' descriptions or else image meanings are indiscernible (Cross, et al., 2006; Gauntlett and Holzwarth, 2006; Morgan, et al., 2009). Conceptualizing drawings as products of the time and space in which they are produced and as being imbued with meanings that are multiple, rather than fixed, can be another way to protect against these misgivings (Rose, 2001; Guillemin and Drew, 2010).

Participant-generated drawing in health geography: a pilot study

With an aim to assess the methodological utility of drawing from a health geography perspective, I undertook a pilot study to empirically evaluate a participant-generated drawing method. My objectives were twofold. First, I sought to gauge the suitability of drawing as a method for understanding men's and women's experiences in gym environments to inform the development of my wider multimethod study. Second, I aimed to appraise the value and limitations of drawing, vis-à-vis traditional qualitative methods, from the perspectives of research participants.

The study

I recruited self-identified gym users via an email LISTSERV within the Queen's University (Canada) community subscribed to by faculty, staff, and graduate students. The study was open for a two-week period in which participants could retrieve, complete and submit a participation packet at any time. Packets contained (1) letters of information and consent forms, (2) drawing activity instructions and accompanying questions, and (3) blank sheets of paper and a pencil. Packets were available for retrieval from two campus locations, and a separate box was available for participants to return completed packets in sealed envelopes. Instructions emphasized that no artistic skill whatsoever was required to participate. Participation was anonymous, and no identifying information was collected. Participants were advised to avoid including identifying information in their drawings, such as drawing a recognizable self-portrait or including the name of their gym. The final sample consisted of six useable responses from three men and three women, ranging in ages from twenties through sixties.[2] Despite this small sample size, as my findings below demonstrate, participants' responses were productive in providing insight into gym experiences and the efficacy of drawing as method.

The drawing method consisted of a four-part self-administered activity. Part 1 collected brief demographic information (age, self-identified gender) and details about gym use (frequency, types of activities). Part 2 asked participants to draw, using the paper and pencil provided, in response to the question: 'How do you feel in the gym?' Part 3 asked participants to describe their drawing in writing. In part 4, participants were asked to reflect on the process of drawing in relation to three questions: (1) 'Would you have answered this question any differently if you had not drawn about it (e.g., if you had only talked or written about it)?' (2) 'Do you think drawing is a helpful method to understand people's experiences in gym environments?' and (3) 'Is there anything you would recommend to change about the drawing exercise?'

I did not undertake a visual analysis of the drawings, but rather adopted an auteur theory approach. I thus treated participants' written responses as textual data, which I sorted into a priori codes corresponding with my study objectives as follows: (1) understanding experiences in gym environments, (2) value added by drawing, and (3) limitations of drawing. Within these themes, I identified similarities and differences across responses.

I preface my results discussion below with one key limitation of this pilot. Given that this was a self-administered exercise, removed from an interview context, I had no way to probe participants about their responses. My data thus consisted only of what participants chose to write, and the level of detail varied

2 One response was submitted without a completed consent form and thus was excluded from the study. This study was granted clearance by the Queen's University General Research Ethics Board according to the recommended principles of Canadian ethics guidelines and Queen's University policies. Participants consented to the publication of their drawings for scholarly purposes. Participants were not offered any compensation.

significantly across responses, with some providing lengthy paragraphs and others including only a few short lines.

Understanding experiences in gym environments

In describing their drawings, participants tended to focus on either articulating immediate bodily sensations of exercising or their relationships with the gym environment and its impacts on their emotional states. This scalar split suggests that drawing can help to illuminate multiple geographies of the gym, from the proximate geography of the body to the interrelationships between individual experience and the socio-spatial context of the gym.

With regard to the physical experience of gym exercise, participants identified sensations of both pleasure and pain. One man in his sixties, for example, who used a gym two to three times per week, cited happiness associated with the feeling of exercising and the sweating that goes along with it, explaining, 'My drawing depicts me – happy to be exercising and sweating – a good thing'. For this participant the external gym environment did not factor explicitly into his representation of his gym experience. A woman in her forties, who used the gym several times per week for individual and group exercise, also represented physical sensations in her drawing. She described that she drew 'a BIRD STUCK ON A CROSS with wings out strengthened, in some pain. I suffer from tightness + soreness in my upper back and shoulders' (capital letters and symbols in original). These responses emphasized the physicality and individuality of some gym experiences.

For other participants, the external atmosphere of the gym factored more prominently into their feelings of well-being and behaviors in the gym, both positively and negatively. For example, a man in his thirties who used the gym one to two times per week for cardiovascular exercise explained that, 'I drew an awkward, off-balance, portly man. He is alone. This is generally how I feel at the gym. Though I am not a "portly" or awkward guy – and I'm fairly active – when I use the gym I feel out of place.' By drawing an intentional misrepresentation of his physical self, this participant expressed the mismatch he experiences between his sense of self and the gym environment. Another participant, a woman in her twenties visiting a gym three to four times per week for cardiovascular exercise and recreational sports, underscored that the gym is not an emotionally homogenous entity, but rather that her experiences vary throughout. Representing this varied emotional terrain in her drawing, she explained:

> Sometimes going to the gym makes me feel very powerful or can fill me with pride, as though I've won a competition with myself and others. . . . In the bottom right corner, I drew a bunch of faces staring at me, as well as a surveillance sign. I often feel as though everyone is watching me – judging me, assessing me – and it makes me uncomfortable and nervous.

In contrast, a man in his twenties using the gym for strength training five to six days per week, described how the gaze of others positively affected his experience.

He enjoyed gaining the attention of nearby exercisers, explaining, 'I drew my friend and I lifting together because when I'm in the gym with her I am generally happy. . . . I also drew some guys who are surprised by how strong she is. This makes me feel unique in the gym because many guys lift with their other male friends.' For this participant, standing out was a positive influence on his feelings in the gym.

These results suggest that drawing yielded insights into the highly heterogeneous and individualized nature of gym experiences. Participants reported multiple scales and layers of feeling in gym environments, spanning the physiological effects of exercise to the psychological impacts of interactions with other exercisers. The responses to drawing were also useful in helping me to critically reflect upon my own assumptions about the nature of the gym spaces, particularly with regard to the extent of their emotionally variegated topography.

Value added by drawing

The majority of participants (five out of six) indicated they would have responded somewhat differently to the question, 'How do you feel in the gym?', if they had instead replied only orally or in writing, without drawing. For two participants, drawing was significant in surfacing new self-understandings that they credited decisively to drawing. For one respondent, the woman in her forties using the gym several times per week, the drawing exercise was instrumental in bringing to the fore aspects of her gym experience that would not have been readily apparent otherwise. She explained: 'I think I might have written about 1 or 2 of the elements I mentioned. But I'm struck that I would NOT have mentioned the tree (and balancing stuff) if I hadn't been drawing the river' (emphasis in original). The tree, as per her description and shown in Figure 8.1, represented feelings of rootedness and peacefulness while exercising in the gym – despite the physical pain she mentioned in her back and shoulders. Similarly, the participant whose drawing depicted him 'happy to be exercising and sweating', shown in Figure 8.2, explained that drawing enabled him to identify aspects of his gym experience that he might not have pinpointed in writing or talk. For him, drawing worked to strip away 'extraneous details that did not capture the "essence" of the exercise experience. Unless asked directly "did exercise in a gym make me feel happy" I might not actually have articulated that.' Drawing for this participant provided a mechanism that helped to reveal aspects of his gym experience were not immediately recognizable to him.

Participants who cited value in drawing indicated that the drawing process was productive in fostering reflection and memory recall in their responses. For the man in his thirties who felt 'out of place' in the gym, drawing served a functional purpose in aiding to concretize memories, but did not substantively affect his self-knowledge or the content of his response. He noted that, without drawing, his response would have been, 'Probably a bit different. Drawing a picture shifts the focus instead of pure memory – I now have a visual to work from. However, the general info I give wouldn't have changed much.' Despite not changing the content of his response per se, this participant still recognized drawing as a beneficial adjunct in research, noting, 'Being able to create is always an asset to

Figure 8.1 Drawing by a participant who credited the process of drawing with the acquisition of new self-knowledge

understanding experiences. It broadens the emotional spectrum.' Similarly, the woman who drew the scene of the tree and river explained that drawing 'slows down the thinking – or maybe more so – intensifies it in a different way. Even though I'm not a great drawer – I felt I could convey something better than I could in words. That "something" was interesting to me as a participant.' For these participants, both the product and process of drawing encouraged thinking about the

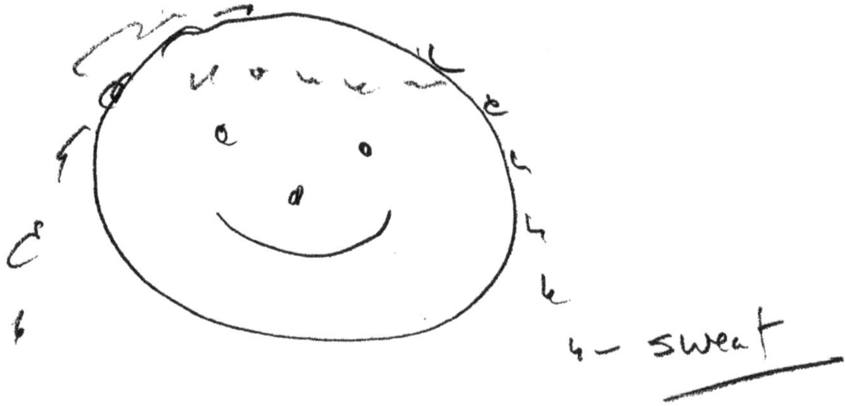

Figure 8.2 Drawing by a participant who credited the process of drawing with helping to pinpoint 'the "essence" of the exercise experience'

research question in ways distinct from purely oral or written responses. Drawing fulfilled this methodological function by providing a visual cue that aided in connecting past memories to the present responses and by opening up possibilities for different thought processes and timetables than those typically associated with traditional qualitative methods. Overall, these responses lend support to the notion that, for some people, drawing can fulfill a distinct methodological function that can complement other qualitative methods in fruitful ways.

In contrast, rather than augmenting their responses, two participants indicated that drawing had a limiting effect for them, which I explore in detail in the section below. Both explained that drawing did not enable them to capture and communicate certain details, such as naming particular emotions, with the level of precision that they would have used if they had responded in oral or written form. The drawing activity had no impact on how one respondent – the man in his twenties using the gym for strength training – answered the question. He cited his own familiarity with the gym setting as a possible explanatory factor.

Overall, from the perspective of participants, the question as to whether drawing *adds* value to qualitative research design was mixed. Half of these participants indicated that drawing positively influenced their research contributions, while two participants found drawing to be constraining, and a third was indifferent. Among the three participants for whom drawing did make a positive difference, the magnitude and nature of this difference varied widely, from unearthing self-knowledge to serving as a representational tool and memory aid.

Limitations of drawing

Two participants indicated that drawing was a constraining means of communication, one citing a preference for written expression and the other pointing to the boundaries of unisensory data. These participants underscored that drawing is not a sensory fit for everyone. For example, one woman in her fifties, who used the gym two to three times a week for cardiovascular exercise and strength training, provided a collage of words instead of an illustration, stating: 'I was unable to draw about my gym experience in the first place. The written word is better for me. . . . When drawing doesn't work for people, like me, there is absolutely nothing there to express that would not be contrived.' Despite being 'unable' to draw, the collage of words itself, shown in Figure 8.3, was very descriptive about the participant's feelings in her gym environment and provided useful insights into her experiences. This suggests that drawing can be operationalized by participants in diversely productive ways, not limited to the pictorial aspects of drawing. Indeed, this response highlights the importance of allowing for flexibility in how participants take up drawing. Another participant, the woman in her twenties, who drew a very detailed depiction of the varied emotional landscape of the gym (Figure 8.4), explained that the visual and written components of the activity curbed the scope of her expression:

> If I could speak, I would offer a more descriptive account of what I think is a very multi-sensory experience (rather than just visual). I might also express tone differently – e.g., I don't know if my frustration came out in one drawing, or if it came across as more angry, confused, etc.

This participant's experience exemplifies the importance of applying drawing as an adjunct to other qualitative methods that allow for talk or text; however, it is not always possible to include all communicative outlets in every study. Still, the limitations of such choices should be critically reflected upon and discussed when using drawing as method.

Conclusions

Throughout this chapter, I have made the case that the processes of drawing have the potential to generate qualitatively different research products, and these present some exciting possibilities for advancing health geography research. Drawing can add to our qualitative palette in health geography by bringing emotions to the surface, supporting a participant-centered research process, and reinforcing rigor. My pilot study supported the methodological utility of drawing, with some participants valuing its reflective function and crediting drawing with the acquisition of new self-knowledge. At the same time, my analysis demonstrates that drawing should be employed critically and flexibly, with attention to the communicative preferences of participants and the limitations inherent in distinct sensory modes of expression. Even when drawing does not 'work' for a particular individual per se, the results can still be significant (cf. Guillemin, 2004). The word collage provided by one of my participants was highly effective in portraying how she felt in the gym. It is quite remarkable that this participant self-selected to engage in this drawing study despite her stated preference for writing. This underscores

Figure 8.3 Drawing by a participant who preferred written over visual expression

Figure 8.4 Drawing by a participant who noted that neither visual nor written representation could capture the tone with which she would have conveyed her experiences in speech

the point that drawing as a method does not need to be pictorial. Indeed, perhaps the method itself requires a renaming. As this chapter has demonstrated, creating representations freehand – in whatever form – can be a productive technique in constructing and representing knowledge in health geography.

Acknowledgements

I gratefully acknowledge the study participants who gave generously of their time and knowledge. Thanks to Mary Louise Adams, who provided me an initial platform to explore drawing as method, to Joyce Davidson and Sarah Turner for their critical comments on earlier drafts, and to the book's editors for valuable feedback which helped me to strengthen this chapter. Funding held during the pilot study and writing of this chapter included an Ontario Graduate Scholarship and a CIHR Doctoral Research Award.

References

Andrews, G., Hall, E., Evans, B., and Colls, R., 2012. Moving beyond walkability: on the potential of health geography. *Social Science and Medicine*, 75(11), pp. 1925–1932.

Azzarito, L., 2010. Ways of seeing the body in kinesiology: a case for visual methodologies. *Quest*, 62(2), pp. 155–170.

Backett-Milburn, K., and McKie, L., 1999. A critical appraisal of the draw and write technique. *Health Education Research*, 14(3), pp. 387–398.

Banks, M., 2008. *Using visual data in qualitative research.* (ebook) London and Thousand Oaks: SAGE Publications. Available through Queen's University Library website: http://library.queensu.ca/. (Accessed November 15, 2012).

Baxter, J., and Eyles, J., 1997. Evaluating qualitative research in social geography: establishing 'rigour' in interview analysis. *Transactions of the Institute of British Geographers*, 22(4), pp. 505–525.

Blacksher, E., and Lovasi, G. S., 2012. Place-focused physical activity research, human agency, and social justice in public health: taking agency seriously in studies of the built environment. *Health and Place*, 18(2), pp. 172–179.

Broadbent, E., Petrie, K. J., Ellis, C. J., Ying, J. and Gamble, G., 2004. A picture of health-myocardial infarction patients' drawings of their hearts and subsequent disability: a longitudinal study. *Journal of Psychosomatic Research*, 57(6), pp. 583–587.

Brown, T., and Duncan, C., 2002. Placing geographies of public health. *Area*, 34(4), pp. 361–369.

Colls, R., and Evans, B., 2013. Making space for fat bodies? A critical account of 'the obesogenic environment'. *Progress in Human Geography*, 37(5), pp. 1–21.

Crawford, R., 1980. Healthism and the medicalization of everyday life. *International Journal of Health Services*, 10(3), pp. 365–388.

Cross, K., Kabel, A., and Lysack, C., 2006. Images of self and spinal cord injury: exploring drawing as a visual method in disability research. *Visual Studies*, 21(2), pp. 183–193.

Davidson, J., and Milligan, C., 2004. Embodying emotion sensing space: introducing emotional geographies. *Social and Cultural Geography*, 5(4), pp. 523–532.

Dworkin, S. L., 2003. A woman's place is in the . . . cardiovascular room? Gender relations, the body, and the gym. In Bolin, A., and Granskog, J. (eds.) *Athletic intruders:*

ethnographic research on women, culture, and exercise. Albany, NY: State University of New York Press. pp. 131–158.

Dyck, I., 1999. Using qualitative methods in medical geography: deconstructive moments in a subdiscipline? *The Professional Geographer*, 51(2), pp. 243–253.

Fenton, N. E., Elliot, S. J., Cicutto, L., Clarke, A. E., Harada, L. and McPhee, E., 2011. Illustrating risk: anaphylaxis through the eyes of the food-allergic child. *Risk Analysis*, 31(1), pp. 171–183.

Gauntlett, D., and Holzwarth, P., 2006. Creative and visual methods for exploring identities. *Visual Studies*, 21(1), pp. 82–91.

Guillemin, M., 2004. Understanding illness: using drawings as a research method. *Qualitative Health Research*, 14(2), pp. 272–289.

Guillemin, M., and Drew, S., 2010. Questions of process in participant-generated visual methodologies. *Visual Studies*, 25(2), pp. 175–188.

Guillemin, M. and Westall, C., 2008. Gaining insight into women's knowing of postnatal depression using drawings. In Liamputtong, P., and Rumbold, J. (eds.) *Knowing differently: arts-based and collaborative research methods*. New York: Nova Science Publishers. pp. 121–139.

Haraway, D., 1988. Situated knowledges: the science question in feminism and the privilege of partial perspective. *Feminist Studies*, 14(3), pp. 575–599.

Harrison, B., 2002. Seeing health and illness worlds-using visual methodologies in a sociology of health and illness: a methodological review. *Sociology of Health and Illness*, 24(6), pp. 856–872.

Johnston, L., 1998. Reading the sexed bodies and spaces of gyms. In Nast, H. J., and Pile, S. (eds.) *Place through the body*. London: Routledge. pp. 244–262.

Kearns, R. A., 1993. Place and health: Towards a reformed medical geography. *The Professional Geographer*, 45(2), pp. 139–147.

Krumpal, I., 2013. Determinants of social desirability bias in sensitive surveys: a literature review. *Quality and Quantity*, 47(4), pp. 2025–2047.

Mannay, D., 2010. Making the familiar strange: can visual research methods render the familiar setting more perceptible? *Qualitative Research*, 10(1), pp. 91–111.

Mays, R. M., Sturm, L. A., Rasche, J. C., Cox, D. S., Cox, A. D. and Zimet, G. D., 2011. Use of drawings to explore US women's perspectives on why people might decline HIV testing. *Health Care for Women International*, 32(4), pp. 328–343.

Morgan, M., McInerney, F., Rumbold, J., and Liamputtong, P., 2009. Drawing the experience of chronic vaginal thrush and complementary and alternative medicine. *International Journal of Social Research Methodology*, 12(2), pp. 127–146.

Nowicka-Sauer, K., 2007. Patients' perspective: lupus in patients' drawings. *Clinical Rheumatology*, 26(9), pp. 1523–1525.

Parsons, J. A., and Boydell, K. M., 2012. Arts-based research and knowledge translation: some key concerns for health-care professionals. *Journal of Interprofessional Care*, 26(3), pp. 170–172.

Reason, M., 2010. Watching dance, drawing the experience and visual knowledge. *Forum for Modern Language Studies*, 46(4), pp. 391–414.

Reynolds, L., Broadbent, E., Ellis, C. J., Gamble, G. and Petrie, K. J., 2007. Patients' drawings illustrate psychological and functional status in heart failure. *Journal of Psychosomatic Research*, 63(5), pp. 525–532.

Rose, G., 1997. Situating knowledges: positionality, reflexivities and other tactics. *Progress in Human Geography*, 21(3), pp. 305–320.

Rose, G., 2001. *Visual methodologies: an introduction to the interpretation of visual materials*. 2nd ed. London: Thousand Oaks: Sage Publications.
Scott, A., 2009. Illness meanings of AIDS among women with HIV: merging immunology and life experience. *Qualitative Health Research*, 19(4), pp. 454–465.
Shinebourne, P., and Smith, J. A., 2011. Images of addiction and recovery: an interpretative phenomenological analysis of the experience of addiction and recovery as expressed in visual images. *Drugs: Education, Prevention and Policy*, 18(5), pp. 313–322.
Thien, D., and Del Casino, V. J., Jr., 2012. (Un)healthy men, masculinities, and the geographies of health. *Annals of the Association of American Geographers*, 102(5), pp. 1146–1156.
Turner, S., and Coen, S. E., 2008. Member checking in human geography: interpreting divergent understandings of performativity in a student space. *Area*, 40(2), pp. 184–193.
West, C., and Zimmerman, D. H., 1987. Doing gender. *Gender and Society*, 1(2), pp. 125–151.

9 Applying decolonizing methodologies in environment-health research

A community-based film project with Anishinabe communities

Chantelle A. M. Richmond

Reflections on praxis

As an Indigenous scholar, one of my primary objectives is to do research that will benefit Indigenous communities. While my lifelong goal is to contribute to the lessening of health and social inequities borne by Indigenous peoples, the truth of the matter is that thinking on such a grand scale can be a heavy burden for a young scholar like me. And if I am honest, such a contribution is likely outside the range of skills and capacities I have to offer. While in Canada we are seeing some improvement in major health indicators, the reality is that the shift needed to begin equalizing Indigenous and non-Indigenous patterns of health is truly momentous. But positive change is possible, and there are plenty of reasons to be hopeful about the contributions that research can make.

I am hopeful, first and foremost because I know this great burden is not mine to carry alone. While the research I do is motivated in many ways by my own personal experiences as an Indigenous woman, I humbly recognize that there are many diverse pieces of the health equity puzzle, the solution to which will come from some combination of skills, resources and approaches. But perhaps the most important piece of the puzzle will be revealed through agreement that Indigenous health equity must be a priority for Canadian researchers. In Canadian universities, over the past twelve or so years, an important shift has been taking place in the ways Aboriginal health and healing are conceptualized, and of the means by which research on these topics ought to be carried out. This dedicated, sustained funding for Aboriginal health training environments led to the mentoring, training and capacity building of hundreds of Aboriginal health trainees in the medical and social sciences. It also led to a generalized recognition that Aboriginal health research must engage in community-centered approaches led by ethically responsive methods. I take great comfort in knowing that I am surrounded by other young scholars who not only share my passions for Aboriginal health, but who bear the skills, knowledge and moral capacity needed to do the work necessary.

Building on the idea of doing research that is morally responsive, I am also hopeful because I know that at least part of the solution to improving Indigenous people's health will come from doing research in a way that is responsive to actual community needs, by placing community ethics at the project's foundation. Too often Indigenous research – particularly that originating in our home discipline of geography – has been undertaken as an academic exercise; careers have been established, findings published, grants awarded, promotion and tenure achieved – but has the community benefited? Not always, and in some rare cases, Indigenous communities have experienced considerable harm as a result of the research done in their communities. Today, we are seeing the development and uptake of Indigenous research approaches that are nurtured through community needs and visions. This hopeful new way of doing research is being initiated through the collaborative synergies of academics and communities – by privileging the voices of those on the ground level, the promise of these partnership-based projects are rooted in research approaches that empower communities to meet their goals of self-determination. To me, this is decolonizing research.

While the investments in Aboriginal health training environments have been considerable, the fact that such health disparities exist tells us there is still a lot of work to do. Another vital part of the solution will come from training the next generation of scholars to do research in this innovative paradigm. How do we go about doing this research? Where do we start? How do we generate the sorts of research questions that will yield impactful results? In applied community research with Indigenous communities, there are no singular answers to these questions as all communities are unique and so too are their research needs. However, there are some culturally appropriate ways to engage in research with Indigenous peoples and communities, and it's these sorts of topics that this chapter discusses. In the past decade or so, a good deal of writing has made the case for the uptake of Indigenous methodologies, decolonizing methodologies or some hybrid of the two, yet the literature is quite sparse in its publication of empirical examples. Through discussion of a community-based film project with my home community, the Ojibways of the Pic River First Nation, and one of our neighboring communities, the Batchewana First Nation of Ojibways, this chapter provides an applied example of a project that employed decolonizing methodology. It is my intention to draw attention to this method as a way forward for other scholars, Indigenous and non-Indigenous alike, who wish to do similar research with Indigenous communities. By doing research in this paradigm, we will contribute to the creation of a new narrative of applied community-based research with Indigenous peoples. And over time, I am hopeful this change in the way we approach Indigenous research will translate into Indigenous health equity.

Introduction

In all Indigenous cultures, daily life and continued well-being have historically depended on a deep knowledge and intimate relatedness to one's local ecosystem, which can be referred to as an Indigenous knowledge (IK) system (Battiste and Henderson, 2000). At the most basic level, it is this knowledge system that has enabled Indigenous people to live and thrive in some of the harshest environments on earth. While there is no singular definition of what an IK system is, it generally refers to the culturally and spiritually based ways Indigenous peoples relate to their local ecosystems and to one another to carry out their way of life (LaDuke, 1994). An IK system generally includes a number of central features, including that it is a dynamic, integrated knowledge that is shared orally, and best learned by practicing or doing with others.

In Indigenous communities around the world, community Elders have traditionally held responsibility for sharing their IK with younger generations, through ceremony and other spiritual and social practices that emphasize the importance of sustaining a balanced relationship with the land. In many communities today, however, there is considerable concern that opportunities for IK transmission are becoming eroded over time as processes of environmental dispossession (e.g., contamination from nearby industry) make it increasingly difficult for Indigenous communities not only to access the resources of their traditional environments, but also to interact on the land in ways that will foster sites for IK transmission. In the face of these threats to land-based Indigenous cultural continuity, many isolated and remote Indigenous communities are searching for innovative, contemporary ways to document and preserve their IK for the benefit of future generations.

This chapter describes the creation of the 2013 documentary *Gifts from the Elders*, which represents the culmination of a five-year community-based research (CBR) project that utilized a decolonizing methodology with youths and Elders from two Anishinabe communities from Ontario, Canada, to document and preserve their local IK. The chapter proceeds in the following sections. The first section discusses the term 'decolonizing methodologies' and identifies the use of film with Indigenous communities as one such approach. The second section introduces Anishinabe experiences of environmental dispossession in Northern Ontario and characterizes the ways this research project was co-developed in response to community needs and desires, including the hiring and training of local youths, as well as community involvement in research planning and data analysis. The third section discusses the research process undertaken in the making of this film. The chapter concludes with a discussion of the transferability and potential application of this project by Indigenous communities in Canada and elsewhere.

Decolonizing methodologies: using community-based film with Indigenous communities

The project described in this chapter was designed jointly between community collaborators from the Ojibways of the Pic River First Nation, the Batchewana First Nation of Ojibways, Turtle Island Productions, and researchers and student trainees from the University of Western Ontario, Lakehead University and Confederation College. Spanning five years (2008–2013), this project demonstrates an applied, CBR project, wherein the methodology grew from the joint ideas, passions, skills and needs of both the community collaborators and our academic team (Tobias, et al., 2013; Castellano, 2004), with particular emphasis on the sharing of knowledge and capacity building (Christensen, 2012;

de Leeuw, et al., 2012). Participation in CBR represents an active means by which Indigenous communities can participate in and shape research that will have direct benefits in their own lives (Louis, 2007; Castleden, et al., 2012). It is a collaborative approach to research which is critical for ensuring benefits for both researchers and the researched, including the ability to share in leadership, decision-making, capacity building and other knowledge and benefits that result from the process (Israel, et al., 2005; Minkler, 2005), including – and perhaps most significantly – demonstrated relevance for local people, and use of this research as a tool for self-determination among Indigenous communities (Coombes, et al., 2014; Smith, 1999; Wallerstein and Duran, 2006). This represents a significant turn from the way environment and health research has historically occurred in the Indigenous context, particularly in the discipline of geography. During the 1990s, for example, several Anishinabe communities on Lake Superior participated in the EAGLE Project (Effects on Aboriginals for the Great Lakes Environment), a four-million-dollar project funded by Health Canada to undertake research examining health-environment links among 37 First Nation communities on the Great Lakes. While the EAGLE Project was successful in obtaining critical baseline health and environment data for the region, there were significant limitations related to the project's narrow definition of health (e.g., mainly biomedical), its limited use of ethnographic inquiry and the failure of the research to integrate local communities and their IK. Unfortunately, the EAGLE Project is not an isolated example of environment and health research gone wrong in the Indigenous context (Deloria, 1995; Smith, 1999; Castellano, 2004; Adams, et al., 2014), and this legacy has left many communities suspicious and mistrusting of researchers.

While we are beginning to make strides in applied CBR with Indigenous communities (Tobias, et al., 2013; Adams, et al., 2014), research in the First Nation context has not adequately developed methodologies capable of measuring the cultural dimensions linking health and environment, and some would argue that the basis of this misunderstanding is related to the way research has been conducted on these topics, that is, as research has too often been framed as an academic exercise (Coombes, et al., 2014), and very rarely as projects designed with the community's needs placed at its center (Castellano, 2004). The research described herein is part of a new and slowly growing research movement called 'decolonizing methodologies', which refers to an approach that places Indigenous communities and their concerns and aspirations at the forefront of research on the health, environmental and social issues that affect them, the objective being to enable communities to use research in ways that will help them to be self-determining (Smith, 1999; Louis, 2007; Kovach, 2009). As a response to the ongoing effects of environmental dispossession and colonialism among Indigenous communities in Canada and around the world – including the ways academic institutions have attenuated these processes – decolonizing methodologies are critical for enabling Indigenous communities to not only be part of research on the important matters that affect them, but to play a central role in guiding the way the research unfolds from the very start. Ideally, such participation is important for ensuring that the project and its subsequent results are impactful and useful at the community level. As we strive to put communities at the center of research agendas, we also struggle to ensure that the appeal and uptake of First Nation's environment and health research is shared in a way that makes it accessible to target populations, policy audiences and the research community. And this is how and why community-based film is such an important methodology and tool for disseminating the stories of Indigenous communities.

Community-based film provides opportunities for self-representation, exploration and alliances amongst Indigenous communities (Dowell, 2006; Mistry and Berardi, 2012). Film allows for a simple and straightforward way to communicate important knowledge to many people at one time. This medium is being used with increasing frequency by Indigenous communities wishing to educate the public about critical issues in their communities and to share their experiences with other Indigenous communities across the globe. Indigenous media is a powerful arena through which communities can document their cultural traditions and knowledges, counter dominant media misrepresentations and tell stories from their own perspectives (Dowell, 2006). Over the last couple of decades Indigenous communities have utilized film as a way to document their local knowledge, to practice their self-determination and to highlight the need for environmental protection and justice (Ginsburg, 2004; Singer, 2001; Iseke, 2011). Canadian examples include Thunderstone Pictures' *Seeking Biimadisiwin*, Alanis Obomsawin's *Kanehsatake: 270 years of resistance*, and the more recent *Qapirangajuq: Inuit knowledge and climate change*, by Zacharias Kunuk and Ian Mauro. Whereby traditional forms of academic research and dissemination often fall short on this criterion, the medium of film – particularly those embedded within a CBR model – can have important transformational outcomes in studying communities (Mistry and Berardi, 2012) and is well suited for widespread knowledge transfer to others (Kindon, et al., 2007; Parr, 2007; Williams, et al., 2003).

In the beginning: detailing Anishinabe struggles to maintain Indigenous knowledge

This research was initiated in 2008, when preliminary meetings were held in several Anishinabe communities across the North Shore of Lake Superior to engage residents, elected officials, band employees, youths and Elders in discussions on their perceptions of local environmental and health issues. These early discussions were fundamental to the substantive and methodological development of the larger research project, and it was these discussions from which the idea of creating a documentary was born.

In the time since the first explorers made their way through the resource rich Great Lakes Region, the Anishinabe people have endured various types of environmentally exploitative resource development, beginning with the fur trade and, in the advent of the industrial revolution, economic development revolving around forestry, mining and steel production. In the past 50 years, these powerful industries have formed the cornerstone of the region's economic profile; however, the economic benefits of these industries have rarely been realized by First Nation communities themselves. Paradoxically, the development of industries in northern Ontario has often led to the contamination of or the complete marginalization of First Nation people from their traditional territories, thereby contributing to a considerable shift in the Anishinabe way of life and the ways in which their IK systems are maintained and practiced. During meetings with the two study communities, these concerns were raised as paramount.

Pic River is located on the north shore of Lake Superior, roughly midway between the cities of Sault Ste. Marie and Thunder Bay, Ontario, and Batchewana is located 20 kilometers east of the city of Sault Ste. Marie, at the Eastern edge of Lake Superior. In both communities, there is considerable angst about historic, present and future industrial development in their traditional territories and the various ways these industries have compromised community ties to the land and their cultural continuities. For example, community members from Pic River explained how a burst tailings line at Hemlo Gold Development (located 40 kilometers upstream) contaminated their drinking water supply, forcing the community to rely on bottled water for several years in the 1990s. Another common narrative from Pic River revolved around the creation of a nearby National Park during the late 1970s, and the limiting impact this has had on the community's ability to practice traditional hunting and harvesting activities within the park boundary, which consumes a considerable portion of their traditional territory. Individuals from Batchewana First Nation spoke with concern about the steel industry in the neighboring city of Sault Ste. Marie. They raised questions about water and air quality, and the safety of consuming fish and mammals from these waterways. Batchewana residents were also expressive about recent wind turbine development on the shore of Lake Superior and the effect of ground vibration for migrating animals, and others who call this area home. While there are economic benefits for those community members who become employed in these industries, for the most part, community members lack the education and training requirements.

Despite the geographic distance between them, and the fact that the processes of environmental dispossession they endure has been different, both Pic River and Batchewana detailed remarkably similar concerns about the consequences of these industrial threats to the health and well-being of their communities and the continued practice of their IK systems. Limited control over and reduced quality of their local environmental resources has led to decreased access to the cultural and social determinants of health, such as participation in hunting, fishing and trapping, and other activities (e.g., cultural practices and ceremonies, harvesting of traditional foods and medicines, opportunities to learn language). In particular, deep concern was expressed by Elders in the two communities about a growing disenchantment of local youths with regard to Anishinabe culture, language and traditional ways of connecting with the land and, on a more pragmatic level, about what this growing disenchantment means for the future preservation of their Indigenous knowledge, local language and culture, and the resolution of community land claims.

These early discussions demonstrated an undisputed need to develop a methodology that would not only lead to the documentation of local experiences and knowledge, but might also re-establish the social connections between youths and Elders, and ignite passion in the youths about their community's environmental futures. In brainstorming the methodology, we drew from an open slate to explore our collective interests and ideas. We were interested in creating opportunities for the youths to become involved in the research because they perceived it was exciting or relevant to them. As a result, most of the ideas revolved around various forms of arts-based inquiry, which we imagined youths would enjoy being part of, including photography, journals, poetry and film.

While these early discussions were important for fleshing out ideas and brainstorming without limitation, it was also a critical time for the research team to be honest about our own skills and capacities (including research funding to carry out said work) so that we would not set up unrealistic expectations. Community-based research requires considerable investments of time and research funding, both which can place limitations on even the best laid plans. In our case, it was only after meeting with several communities, all of whom supported the idea of a film, that we engaged with a First Nation film-maker and took the steps necessary to secure the funding for the project. And even then, it took two years before we were successful in securing the research funds necessary to produce the film.

Submitted in winter 2010, our grant proposal detailed a study that would hire and train local First Nation youths to interview community Elders about health and environmental issues – with emphasis placed on preserving IK – while at the same time providing important opportunities for capacity building and youth empowerment. By detailing this process through a documentary film, the greater objective was to create a lasting memory of the IK bases of the two communities. The media of film was agreed on as a culturally meaningful, high impact method to capture IK and share it broadly in the local community (as a method of knowledge transfer between Elders and youths), as well as to the global Indigenous community so that it might provide a template for other communities wishing to undertake similar projects.

Negotiating the research plan

Meaningful consultation, participation and relationship building are crucial components of the development and consent process for any Indigenous CBR project (Minkler, 2005). Considerable time and effort was placed on initiating these processes long before we secured the research funding, and this relationship building continued as the research plan was actualized. During the first six months of this project, we focused on two specific tasks. The first task was to work with community collaborators to set up local advisory committees (LACs) in both Batchewana and Pic River. The LACs were represented by various segments of the community, including Elders, youths, elected and non-elected officials and band employees, and their roles were to facilitate community meetings and maintain the communication lines between the university researchers and community collaborators.

During previous community visits, many individuals indicated willingness to participate in such a way, and these activities were laid out in our community research agreements. The second task focused on the recruitment of three youths (less than 30 years of age, full time post-secondary students) from the two communities (for a total of six), who would be trained and hired to conduct interviews with the Elders for the documentary. Additionally, our film-maker worked with our co-investigator at Lakehead University to hire two First Nation film-studies students from Confederation College, located in Thunder Bay, Ontario, who would work alongside the youth interns and graduate students to film the interviews.

While our greater objective was focused on the preservation of IK, we recognized that we could not proceed without significant attention paid to training and capacity building for our youth interns and film studies students. In order to provide them the skills needed to undertake their summer employment, in the week leading up to the start of their paid internships, we held a week-long summer school at Western University, in London, Ontario. The purpose of the Summer School was primarily to get the entire team on the same page, and secondarily to provide the youths and film studies students with the theoretical and practical research skills needed to undertake the interviews with the Elders in their communities. This summer school brought together the six youth interns, two film studies students, community collaborators from Pic River and Batchewana (including one Elder), three graduate students, our film-maker and investigators from Lakehead and Western University. The summer school content was provided through workshops and participatory learning techniques on a range of topics, including qualitative research, researcher positionality, ethical principles, interview skills and use of technology in research (e.g., digital recorders, film). The intent of having the summer school at Western University – as opposed to one of the communities (which might seem more appropriate for a community-based project such as this) was to enable the youths to experience the university context. While the visit to Western meant traveling by plane – the first time for some – and time away from family and friends, ongoing research with First Nation youths from Northern Ontario indicates that visiting the post-secondary institution is the most effective way to recruit students to the university. The post-secondary context is an unfamiliar environment for many First Nation communities. We wanted to empower the youths with a positive experience so that they could visualize themselves returning. Ultimately, we intended to provide an opportunity where the youths could gain the skills and expertise needed to successfully complete the responsibility of their summer internships, but also to see how post-secondary training can be used in ways that align with their local community contexts and needs.

When organizing a complex CBR project involving multiple communities, investigators and strategic partners, it is necessary to spend substantial time developing a research strategy that will most efficiently utilize human and material resources. This was particularly important for this project, as all of the stakeholders came in with differing skill levels, expertise and visions for the project. Our film studies students were hired to film the Elder interviews in the two communities.

With their home base in Thunder Bay, Ontario – which is 300 kilometers from Pic River and 700 kilometers from Batchewana – it was critical to develop a finely detailed research plan that would minimize equipment costs and travel time and would take full advantage of time spent with Elders to ensure maximum sampling from both communities. The final task of the summer school was to create a master research plan that detailed two separate weeks of filming in each community, for a total of four weeks of filming. At the end of the summer school, the youths and community collaborators returned to their communities with the graduate students to undertake the filming process.

The making of the film

In order to capture the youths' perceptions about their roles in the research process, and to document their acquisition of knowledge and capacity building during the summer internships, the six youths were also themselves interviewed on film by a graduate student trainee at the outset of the summer school (i.e., while at Western University), and again at the finale of the summer internship. In the first interview, the youths were questioned about key health and environmental issues in their communities, including their ideas for strategies that might improve current realities. The second interview incorporated additional questions to allow for an evaluation by the youths of their knowledge uptake and increased capacity as a result of their participation in the research process.

During their eight week long internships, the youths completed interviews with approximately 23 elders in each of their respective communities (a total of 46 interviews). To ensure ongoing support and feedback to our interns, all interviews were undertaken with the assistance of graduate student trainees (one master's and one PhD student), who resided in each of the two communities for the summer and worked on a daily basis with the interns. The pairing of interns and trainees was meant to provide the interns with some structured support during their internships, and maintain data quality in the interview processes. A long 'roster' of potential Elder interviewees was developed during the summer school. Upon returning home, the interns visited with all potential interviewees to gauge interest and schedule interview times with them. It is important to note that while many Elders consented to have their interview filmed and included in the documentary, some Elders were only interested in having their interviews audio-recorded. Hence, some interviews were scheduled to coincide with the visits by the film studies students, and some were not.

The Elder interviews occurred in various places within the traditional territories of the two communities. Some interviews took place indoors, but most happened outdoors. On average, the interviews lasted two hours, however some went on for up to three hours. Most interviews happened one-on-one, but there were two sets of married couples from each community who co-conducted their interviews. As part of their interviews, many Elders wished to share special places, songs, language and medicines with the youths. While most of the interviews were conducted using a prepared interview guide (which was co-developed by the youths during the summer school), there were some cases wherein Elders spent the duration of the meeting time telling stories, without a single question from the youths, yet still managing to address all the questions in the interview guide. The youths felt they learned the most from these interviews.

The final week of the internship was spent in an end-of-summer school, wherein our graduate student trainees led an intensive four-day, end-of-summer school at Pic River First Nation for interns and collaborators from the two communities. The purpose of this workshop was to enable the youth interns and wider study team to come together to discuss their individual experiences over the summer and to discuss similarities and differences in the key themes resulting from the Elder interviews both within and across the two communities. Another goal of this research was to develop a preliminary analytical framework that could be used for development of the documentary storyline. This process, of co-developing the analytical framework was essential, as this is the stage of CBR where the community typically often falls out of the process (Israel, et al., 2005; Minkler, 2005). Our continued attention to balancing academic and community interpretations of the data meant that this ongoing co-creation was vital to the successful implementation of our CBR approach, as we wanted to make sure that the resulting narrative shared in the documentary aligned as closely as possible to the community's intent. By working with the youths to explore the Elder interviews in a really intensive way, we were able to ensure that the youths' voices and perspectives resonated in an authentic, meaningful way in the subsequent film (Iseke, 2011). It was during this end-of-summer school wherein the second set of interviews was completed with the youth interns, the purpose being to allow for an evaluation by the youths of their participation in the research process itself – including any shift in knowledge base and attitude toward health and environmental issues – and suggestions for addressing these issues. The findings of this research are reported elsewhere (Kulmann, 2012).

The transformation of the filmed interviews into the resulting documentary was a long, arduous and highly collaborative process. Once all interviews were completed, they were transcribed verbatim. All interviews contained the time-stamp information so that the various research team members who undertook analysis of the interview data could easily shift between the transcribed data and the filmed interviews. During the fall of 2011 and winter of 2012, the film director and producer worked in tandem to create a storyline that resonated around the key themes shared in the interviews. At the same time, a number of research assistants were undertaking thematic coding of the interview data, thereby identifying the key stories and text that would form the basis of the film. Beyond the filmed interviews, there are a number of other materials necessary for the creation of a documentary, including b-roll footage (this refers to supplemental film footage that is intercut with the main footage in a documentary, for example, scenic shots that demonstrate the topic an interviewee is discussing, such as a lake, a river or an industry), narration, music, still photos, community profile information (including demographic, geographic and health data) and archived images and film from Northern Ontario and northern Minnesota. The background research and time spent gathering these extra materials was immense and required nearly daily communication between all members of the research team, most specifically the director and producer, who made a number of creative decisions around the placement of these materials in the story to pursue the greatest impact. In February 2012, we achieved the first rough cut of the film, at 100 minutes. This version of the film was shared with the wider research team, including local youths and collaborators. Several edits were subsequently made to shorten the film and to strengthen the threading and impact of the story.

In August 2012, the study team held Elder's gatherings in the two participating communities so that we could disseminate the results from the youth and Elder interviews, and share a screening of the 2nd rough cut of the film. As with all previous stages of this research, the success of the Elder's gathering relied heavily on the direction and contribution of our LACs, as well as the political and spiritual leadership from each community. Each of the gatherings began with a sunrise ceremony, wherein all pipe carriers and bundle holders were invited to participate. The fact that we started our gatherings in this way demonstrates the deep spiritual and cultural significance of the research. During these gatherings, we worked tactfully to ensure the sacredness of these spaces as a place where all attendees would feel safe knowing the knowledge, stories, ideas and perspectives shared would be given the utmost respect. These gatherings were meant to provide a forum wherein all community Elders were invited to discuss their shared concerns about health and environment in their communities, including the development of strategies for environmental repossession (Tobias, et al., forthcoming). The gatherings in both communities were very well attended, and we received constructive criticism from gathering participants on both the results of the interviews and the film. In fact, the viewing of the film led many Elders to express deep emotion and pride in their communities, and great hopefulness that this project would generate interest among youths in their communities to become more involved in traditional activities and to see the benefit of connecting with Elders in ways to improve their IK base.

Following the Elder's gatherings, we integrated the comments and direction received during the community visits into the final cut of the film, which was finalized in Spring 2013. At the same time, we developed a companion website for the film (giftsfromtheelders.ca), which was launched in June 2013, when we held a number of community screenings of the film across Ontario. The resulting film, *Gifts from the Elders*, is a 60-minute documentary that explores the transformative journeys of Anishinabe youths as they spend a summer talking with their community Elders. The film shares Anishinabe teachings, documents special places on the land, and explores processes of environmental dispossession and their impacts on health, well-being and cultural identity of the Anishinabe people. The film concludes with statements by the Elders of their visions for their communities and testimonial by the youths about what they learned as a result of their participation in the research and, perhaps more importantly, of the transformative ways they want to share their knowledge and experiences for the betterment of their communities. The film can be viewed in its entirety on the companion website. Also on this website are biographies of the research team, including the summer interns, film studies students, the graduate trainees, film-makers and academic leaders.

Conclusion

The *Gifts from the Elders* project was nested in a critical new paradigm that is currently framing the way Indigenous health research is being pursued, by providing a flexible, inclusive, interdisciplinary approach that empowers Indigenous community participation throughout the research process and obligates that research outcomes must meet community needs and desires (Wallerstein and Duran, 2006). The project detailed in this chapter and the resulting documentary film represent the application of a decolonizing methodology with Anishinabe communities to address Elder concerns about the preservation of their IK, and create a space for youths to engage meaningfully in research that affects not only their daily lives, but their own cultural continuities and environmental futures. This project pioneered a critically important area of Indigenous research by documenting, on film, the preservation of IK and, in so doing, enabling local youths and Elders to be centrally involved in this process.

In this particular project, the idea of creating a film did not emerge as a result of any one individual's skill set or desires, but rather through a process of collective 'daydreaming' about what the most innovative, exciting and transformative output could look like. In the application of a decolonizing methodological framework, such as that presented here in this chapter, it is important to note there are no hard and fast rules about the ways in which the work ought to be carried out or documented. Ultimately, the ways in which these projects unfold will be as varied as are the communities and their interests. In our project, the idea of producing a film was one that developed organically over time. This entailed a collective discussion about what the communities envisioned this research could become, balanced with frank discussions about skills, expertise, research funding and time needed to make this vision a reality. In this context, the creation of community research agreements that detailed both community and researcher commitments and responsibilities were important for ensuring that both the work and reward was shared in an equitable way. In applied CBR, the best projects will develop when communities and researchers take the time to listen to one another and take stock of the various skills, capacities and resources that can be amassed through such a partnership.

Inspired by Linda Smith's (1999) seminal writings on decolonizing research, this research determinedly engaged the two collaborating communities, as deeply as they wished, in all stages of the research, the intent being to use research as a way of building local capacity and creating space for social action, self-determination and transformation. Perhaps most importantly, this research enabled the preservation of sacred IK, which is being threatened by industrial development and various other processes that dispossess and marginalize Indigenous peoples – in Canada and elsewhere – from their traditional lands and resources. The approach taken in this research may provide a template for other Indigenous communities to become involved in research on issues of local importance, and to use the findings in ways that will strengthen community capacities and idealize what their visions of environmental repossession look like.

In the context of unprecedented environmental change, the uptake of decolonizing methodologies – however they are applied at the community level – are critical for stabilizing IK and inspiring Indigenous communities to engage in research so they may be resilient in their struggles with environment and health issues, and become co-creators of strategies that will enable them to resist and overcome these struggles. By drawing from such an approach, Indigenous communities in Canada, and elsewhere, may be encouraged to become active participants in research that will aid their own community determination of health, strengthen their cultural continuities and protect their environmental futures.

References

Adams, M., Carpenter, J., Housty, J., Neasloss, D., Paquet, P., Walkus, J., and Darimont, C. T., 2014. Towards increased engagement between academic and indigenous community partners in ecological research. *Ecology and Society* 19(3), p. 5.

Battiste, M., and Henderson, J., 2000. *Protecting indigenous knowledge and heritage: a global challenge*. Saskatoon Canada: Purich Publishing.

Brant Castellano, M., 2004. Ethics of Aboriginal research. *Journal of Aboriginal Health* 1(1), p. 98.

Castleden, H., Morgan, V. S., and Lamb, C., 2012. 'I spent the first year drinking tea:' Exploring Canadian university researchers' perspectives on community-based participatory research involving Indigenous peoples. *The Canadian Geographer*, 56, pp. 160–179.

Christensen, J., 2012. Telling stories: exploring research storytelling as a meaningful approach to knowledge mobilization with Indigenous research collaborators and diverse audiences in community-based participatory research. *Canadian Geographer*, 56, pp. 231–242.

Coombes, B., Johnson, J. T., and Howitt, R., 2014. Indigenous geographies III: methodological innovation and the unsettling of participatory research. *Progress in Human Geography*, 38(6), pp. 845–854.

de Leeuw, S., Cameron, E. S., and Greenwood, M. L., 2012. Participatory and community-based research, Indigenous geographies, and the spaces of friendship: a critical engagement. *The Canadian Geographer*, 562, pp. 180–194.

Deloria, V., 1994. *God is red: a native view of religion*. Golden, Colorado: North American Press.

Dowell, K., 2006. Indigenous media gone global: strengthening Indigenous identity on- and offscreen at the First Nations/First Features Film Showcase. *American Anthropologist*, 108(2), pp. 376–383.

Ginsburg, F., 2004. Shooting back: from ethnographic film to Indigenous production/ethnography of media. In Millerr, T. and Stam, R. (eds.) *A companion to film theory*. Malden, MA: Blackwell. pp. 295–322.

Iseke, J. M., 2011. Indigenous digital storytelling in video: witnessing with Alma Desjarlais, *Equity and Excellence in Education*, 44(3), pp. 311–329.

Israel, B., Eng, E., Schulz, A., and Parker, E., 2005. *Methods in community-based participatory research for health*. San Francisco: Jossey-Bass.

Kindon, S., Pain, R., and Kesby, M., 2007. *Participatory action research approaches and methods: connecting people, participation and place*. Abingdon: Routledge.

Kovach, M., 2009. *Indigenous methodologies: characteristics, conversations, and contexts*. Toronto: University of Toronto Press.

Kulmann, K., 2012. We should be listening to our Elders: evaluation of transfer of Indigenous knowledge between Anishinabe youth and Elders. Unpublished master's thesis. Department of Geography, Western University, London, Canada.

LaDuke, W., 1999. All our relations: native struggles for land and life. Cambridge, MA: South End Press.

Louis, R. P., 2007. Can you hear us now? Voices from the margin: using indigenous methodologies in geographic research *Geographical Research*, 45, pp. 130–139.

Minkler, M., 2005. Community-based research partnerships: challenges and opportunities. *Journal of Urban Health*, 82(2), S2, ii3–ii12.

Mistry, J., and Berardi, A., 2012. The challenges and opportunities of participatory video in geographical research: exploring collaboration with indigenous communities in the North Rupununi, Guyana. *Area*, 44(1), pp. 110–116.

Parr, H., 2007. Collaborative film-making as process, method and text in mental health research. *Cultural Geographies*, 14, pp. 114–138.

Richmond, C. A. M., and Ross, N. A., 2008. The determinants of First Nation and Inuit health: a critical population health approach. *Health and Place*, 15(2), pp. 403–411.

Singer, B. R., 2001. *Wiping the war paint off the lens: Native American film and video*. Minneapolis: University of Minnesota Press.

Smith, L., 1999. *Decolonizing methodologies: research and Indigenous peoples*. London, UK: Zed Books.

Tobias, J. K., Richmond, C. A. M., and Luginaah, I., 2013. Community-based participatory research (CBPR) with indigenous communities: producing respectful and reciprocal research. *Journal of Empirical Research on Human Research Ethics*, 8, pp. 129–140.

Wallerstein, N. B., and Duran, B., 2006. Using community-based participatory research to address health disparities. *Health Promotion Practice*, 7(3), pp. 313–323.

Williams, L., Labonte, R., and O'Brien, M., 2003. Empowering social action through narratives of identity and culture. *Health Promotion International*, 18(1), pp. 33–40.

10 Not another interview!

Using photovoice and digital stories as props in participatory health geography research

Heather Castleden, Vanessa Sloan Morgan and Aaron Franks

Introduction

Communities of all sorts, although especially non-dominant communities (e.g. Indigenous communities, communities of colour, LGBTTQTS+ communities and stigmatized communities, such as those who are housing the insecure and homeless, who are living with HIV/AIDS, living with addictions, sex workers, etc.), have said time and time again to academics in the health and social sciences: 'Not another interview! We've been researched to death' (Schnarch, 2004; Jacklin and Kinoshameg, 2008; Brant Castellano and Reading, 2010). Indeed, the interview is probably the most frequently used tool in the qualitative health researcher's toolbox. But there is very good reason; interviews, especially semi-structured and unstructured interviews, create space for interviewers and interviewees to explore a particular topic in great depth, exponentially more than can be achieved through standardized quantitative tools such as survey research. At the same time, interviews can have a 'clinical' feel to them, since they remain under the purview of an interviewer's prerogative as to the direction that they go in. Thus, as many of these non-dominant communities are now focusing on 'researching back to life', especially among Indigenous populations (LaFrance and Crazy Bull, 2009; Castellano, 2014), putting prospective participants at ease and creating more space for autonomy and control over the research process have become important considerations in community-driven participatory health research. We are now seeing more frequent use of 'props' (research tools that act as documentary aids and creative prompts) that can be used to inject a more engaged and reflexive approach which increases the creative and intellectual control of participants in research processes. In this chapter, we illuminate the utility of using innovative props, specifically photovoice and digital storytelling, as methodological tools for engaging in research, especially community-based participatory research (CBPR) in health geography; we also mine our own experiences to offer some suggested dos and don'ts.

We begin by offering preliminary reflections on praxis – thoughts that we feel readers will benefit from carrying with them as they read this chapter. Situating the CBPR approach to research in the early social and political movements of the twentieth century as well as academic researchers' early use of visual methods follows.

We then shift our attention to the cultural turn in health geography in the latter part of the last century, drawing on post-structural, feminist and anticolonial scholarship to explore the emergence of CBPR from within the academic realm. We go on to outline what distinguishes this approach, philosophically and methodologically, from conventional health research. From there, we turn our attention to photovoice and digital storytelling, which are two props in qualitative health geography that are showing promise in terms of positive (i.e., effective) social change and health equity, while embodying key tenets of ethical health research with community partners. Each prop is described in detail for how it might be used in data collection or knowledge mobilization, with a focus on the often blurred differentiation of data and knowledge products, as well as the ethical implications of using audiovisual material that challenge conventional ethical protocols. We conclude the chapter with a second reflection on praxis based upon observations and discussions that we have had, both collectively and independently, over a number of years. We draw together five common themes about the utility of photovoice and digital storytelling to encourage deeper reflection by those considering these props in their own research.

Reflection on praxis (part 1)

From our philosophical stance as co-learners in research, we have been dedicated to decentering the 'enterprise' – from being the sheltered product of academic proprietorship to being in, with and for communities. The props we discuss here as well as other arts-based props have the potential to contribute to this co-creation of new knowledge. With respect to photovoice, looking at other peoples' pictures, sitting at kitchen tables or in living rooms, and hearing stories about what inspired the click of the camera is an honor and a privilege that goes far beyond what the standard operating procedures of the conventional research enterprise can offer. Using photography in research, whether it is participant-employed or researcher-generated, and whether stories about the photos come from participants or from researchers' interpretations about the photos, will illicit new – and different – knowledge. The question to keep in mind as you read this chapter and implement your own projects is: to what end?

Digital storytelling also has a unique capacity to give vision and voice to everyday experiences. Again, with the mindset of decentering research, we were initially attracted to digital storytelling for its potential to keep story makers' voices intact (Lambert, 2013), and we discovered that the opportunities for broader engagement and relevance afforded by this lively medium exceeded our expectations. At the same time, using either as a prop does present risks that can be found in many qualitative forms of inquiry, such as misrepresentation and misappropriation by researchers or people seen as 'outside' of the population creating the stories. Consequences of misappropriation and mishandling of stories in these contexts (whether props are used or not) are severe – story makers may feel silenced, audiences may interpret key themes incorrectly, or research relationships may be wounded – for both the academic and community members alike.

To reduce the risk, co-analysis of digital stories is not only appropriate, but necessary to ensure rigor in research findings, a topic that will be discussed further below.

Photovoice and digital storytelling are not intended to comprehensively reproduce an experience, but rather to provide a snapshot in time (past, present, future; imagined or real-time), to encourage deep thinking and reflexivity, and to weave together the creative visions of the story maker. And while not wishing to discourage the use of either, but particularly digital storytelling, our experiences have shown that unforeseen circumstances, whether that be technological malfunction, story makers' desire to change story topics or decisions to speak in a collective versus individual voice, arise during the process. Researchers would thus do well to reflect on their own motivations for using props, and confer with the people whom they will engage with long before committing to these tools. Some questions that we encourage researchers to carry with them while reading this chapter include, 'Is this prop appropriate given the potential cultural and/or political context?', 'Will participants be interested or disengaged, comfortable or estranged from this prop?' 'Will participants have the time to dedicate to using this prop?', 'How will the prop 'product' be handled afterwards?', 'Does your team have the skills and time available to troubleshoot when using these props?', and 'How will your team navigate the tensions that arise between institutional requirements for research ethics and participants' rights to autonomy for their voice and visions?'

Definition and roots of CBPR

CBPR is an umbrella concept that includes research conducted under many different designations, including action research, participatory research, participatory action research and collaborative inquiry (Kauper-Brown and Seifer, 2006). Scholars sometimes even use these terms interchangeably because the concepts share underlying goals of social change (Minkler and Wallerstein, 2003). The roots of CBPR lie in the social and political movements of the twentieth century. Kurt Lewin (1946), an early innovator in social, organizational and applied research theory, first proposed the concept in the 1940s as a way of confronting issues of social justice and challenging researcher 'objectivity' (Fals-Borda and Rahman, 1991). He introduced the term 'action research', a research process where 'theory would be developed and tested by practical interventions and action; that there would be consistency between project means and desired ends, and that ends and means were grounded in guidelines established by the host community' (Stull and Schensul, 1987, cited in Fox, 2003, p. 88). Paulo Freire (1970), a Brazilian educator and leading philosopher of critical pedagogy, built on Lewin's ideas with the concept of research and education for a critical consciousness, which emphasizes community-based identification of problems and solutions (Tandon, 2002). These consciousness-raising practices, or conscientization, connected strongly with both emerging anticolonial and decolonization movements around the world (Carroll, 2004, p. 276). Current CBPR approaches are indebted to as well as contribute to both anti-colonial and feminist approaches (Kindon, et al., 2007).

Differing from conventional research, CBPR practitioners attempt to *equitably* involve community partners in research; CBPR is intended to honor and draw upon non-academic knowledge and experience, share or give over decision-making responsibilities and build community (Minkler and Wallerstein, 2003) and academic capacity for doing research 'in a good way' (Ball and Janyst, 2008). CBPR practitioners generally recognize research as an inherently political activity and require that researchers adopt a participatory perspective, which 'asks us to be both situated and reflexive, to be explicit about the perspective from which knowledge is created, to see inquiry as a process of coming to know, and to serve the democratic, practical ethos of action research' (Reason and Bradbury, 2006, p. 7, cited in Kindon, et al., 2007, p. 11). CBPR values the process as much as the product – if not more so; in short a 'successful' CBPR project can demonstrate rigorous findings as well as new skills and capacities amongst those involved in the experience (Kindon, et al., 2007). There are thus additional criteria for validity beyond conventional research where rigor is the primary concern to include an evaluation of whether power was equitably distributed (Carroll, 2004). In theory, CBPR levels the playing field, but it is not without its own tensions; there is still an inherently unequal relationship between researchers and research participants even when they are called 'partners' (de Leeuw, et al., 2012).

CBPR is not a method itself, rather it is an approach to research that employs a broad spectrum of techniques that typically involves some form of reflection, dialogue, and action (Kirby and McKenna, 1989). The methods that CBPR practitioners typically employ are those that best facilitate cycles of reflection, research and action in the specific context of the community where the research is taking place (Kindon, et al., 2007). Thus, while there is no single set of methods appropriate to CBPR, the most common methods are those that have traditionally focused on dialogue, storytelling and collective action (Kindon, et al., 2007). Innovation in the use of participant-employed photography, video and computer software, with their 'hands on' qualities, has crept into the repertoire of tools used in CBPR. In short, there is a rising popularity in arts-based techniques that emphasize the co-creation of knowledge through collaborative data collection and analysis. Photovoice and digital storytelling have both emerged as prominent methods that fit this exploratory aim.

Story, storytelling, and narrative

Both photovoice and digital storytelling are concerned with subjective understandings of the world, and as such, work through the related but distinct frames of *narrative* and *story*. With a capacity to communicate memory and spatiality, especially in emotional and feminist geographies, the increasing justification for storytelling as a methodology during the 'cultural turn' of the 1970s leaned heavily upon the work of Foucault, in particular, employing the ability for discourse to be deconstructed as a product of but also constitutive of broader social structures (Razack, 1993). In contrast, a narrative approach more appropriately fits geography's quest for rigor, especially during the long-standing 'qualitative versus quantitative' debates, as literary critiques had developed narrative inquiry as an academic practice (Price, 2010). Remnants of this mentality remain today as narrative is still often privileged over (typically oral) stories, reproducing what Price, in her commentary on cultural geography, identifies as an 'implicit hierarchy' (Price, 2010, p. 204). Knowledge derived from stories, especially in the context of Indigenous-settler colonial contexts like Canada (but also Australia, New Zealand, the United States and elsewhere), has thus been subject to a 'colonizing gesture' (Price, 2010, p. 204), even being misused by researchers through inappropriate handling, analysis and ignorance surrounding protocols of receiving, sharing and interpreting stories (Keeshig–Tobias, 1997; Pittaway, et al., 2010). Appropriate use and engagement with stories is thus especially pertinent in health-focused CBPR involving non-dominant populations.

While definitions for story and narrative abound, Cameron, a historical geographer, suggests that 'what [they] share, most broadly, is a longstanding concern with the ways in which personal experience and expression interweave with the social, structural, or ideological' (Cameron, 2012, p. 2). However, how health geographers and scholars beyond the discipline differentiate between narrative, story and even discourse is a point of inquiry not likely to reach consensus. Cameron suggests that 'geographers working through the relations between story, narrative, and discourse would do well . . . to make their understandings of these concepts and their interrelationships more explicit' (Cameron, 2012, p. 15). Despite disagreement in definitions, the utility of storytelling to contribute to deeper and affective understandings of broad social processes, how these processes are experienced and how they are spatially embedded is largely agreed upon. When viewed in light of increased accessibility to innovative technology, novel uses of storytelling as a tool in research are expanding, not just as a knowledge mobilization conduit but also as a method to collect data. The emergence of photovoice and digital storytelling are two such examples of storytelling and its hybridization with accompanying arts-based technology.

Photovoice

As conceived by Wang and Burris (1997), photovoice uses participants' self-selected photographs as the visual 'database' to facilitate further refection and discussion. The typical procedure for photovoice involves a researcher providing cameras to a group of participants, outlining instruction in their use and information about the ethics and power of photography, and then providing participants with a photography assignment on a particular topic, leaving the photographer/participant to determine what is 'worth' photographing and storying. This process is intended to ultimately engage participants (those typically with less power to influence decision-making) and policymakers (those typically with more power) in a group dialogue with the intent for social change (Wang and Burris, 1997). Wang (1999) identifies three major theoretical underpinnings associated with the development of photovoice. The first, documentary photography, is premised upon providing a camera to people who might not normally have access to one, which will empower them to record and instigate change in their communities (Rose, 1997). The second is Paulo Freire's (1970) theory of critical consciousness, which seeks to engage individuals in the questioning of their historical-social situation. The third, feminist theory, meant for non-dominant populations, values knowledge grounded in experience, takes into account patriarchal power and representation and recognizes local expertise and insight that cannot be fully realized from the outside (Hesse-Biber and Yaiser, 2004). In keeping with the aims of traditional CBPR, Wang and Burris (1997) identify three goals for photovoice: (1) to assist individuals with recording and reflecting on select community issues, (2) to encourage group dialogue on these issues, and (3) to influence policymakers.

The original photovoice model follows several linear steps and is designed for simplicity, ease of use, and transferability. The six key stages below are adapted from Wang (1999, p. 187–189):

1 Select a target audience (e.g., policymakers or community leaders) and identify who has the power to make decisions that can improve the situation in question. Ideally this group will serve as an ad hoc advisory board with the political will to put participants' ideas into practice.
2 Recruit project participants. Seven to ten people are suggested as ideal for balancing logistics and in-depth discussion.
3 Hold a first meeting with participants. Familiarize participants with the project aims and basic ethical principles of photography, explain and obtain informed consent, pose the initial theme(s) for taking pictures to be discussed and agreed upon by the participants group and facilitator), and distribute cameras and instruction on their use.
4 Provide a fixed time frame for participants to take pictures. One week before participants turn in their film or digital photos, and another week before meeting to discuss their photo are suggested.
5 Hold a second meeting with participants to co-analyze photographs.
    a Participants select and discuss the photographs they feel are most significant or meaningful.
    b Participants share stories about their photographs, which serves as additional data and co-analysis.
    c Participants identity the key issues, themes or theories from their discussion about their photographs.
6 Plan a format to share photographs and stories with target audience (e.g. gallery exhibit, online, coffee table book, etc.).[1]

1 For a detailed guide to photovoice, the Prairie Women's Health Centre of Excellence (2009) developed an excellent guide with practical tips and a checklist for planning photovoice projects: http://www.pwhce.ca/photovoice/pdf/Photovoice_Manual.pdf.

Variations to this model have since taken many forms. For example, in my earlier work (Castleden, et al., 2008), our team's community advisory committee suggested we provide the opportunity for participants to discuss all of their photographs with a member of the research team individually instead of in a group setting to allow for more freedom of expression. Participants then selected which photographs and stories they wanted prioritized on a thematic three by five foot poster, which was subsequently printed and displayed at a monthly 'poster release party' and potluck dinner to showcase emergent findings. Doing this also created an ongoing cycle of recruitment, wherein new participants emerged in response to emergent findings, whether it was to support the findings or to identify other community-derived priorities. This also led to some 'completed' participants requesting a second camera to continue this iterative process. Thus, rather than it being a one-to-two week process, the data collection and co-analysis continued for six months, allowing for multi-vocality on the environment and health issues under study.

While any single photovoice project might employ multiple innovations, a review of the literature points to three overarching trajectories. First, as noted above, many researchers employ photovoice in the context of CBPR, and thus there is often a need to lengthen or add stages to the prop, or embed the prop in a larger project that also uses other methods and data sources. For example, Pigford and colleagues' (2013) use of photovoice was embedded in a larger community health project on Type 2 diabetes that also employed quantitative data, asset mapping and interviews with children. Jones and colleagues (2010) conducted four semi-structured interviews with parents over the course of a year in addition to using participant photography in order to develop relationships with extended multigenerational Māori families while tracking seasonal asthma variations. Second, how photographs are analyzed is also being approached with novelty. For example, in Mejia and colleagues (2013) work with Latina mothers, participants discussed and critically analyzed all of their photographs rather than selecting one or two for particular attention. This was done to foster opportunities for participant storytellers to be *pensadoras* (creative thinkers) by weaving as many images as they liked into their personal narratives. In order to allow participants to critically enhance their focus throughout Valiquette-Tessier and colleagues's (2015) study, participating low-income single mothers repeated stages 4 and 5 (taking, sharing and analyzing photos), resulting in a second iteration of relevant (or in Freireian terms, 'concrete' [1970]) themes. Third and finally, photovoice literature sees innovation in the area of materials additional to participants' photographs brought into the discussion and thematization stage (stage 5). There are many reasons for including materials other than participants' photos in the analysis stage. Often participants will be attracted to alternative visual formats such as drawing (Jones, et al., 2010; Young, et al., 2013), or in the case of Shea and colleagues' (2013) work with young Indigenous women and their body image, collages using popular magazines. In their work with

African Americans managing chronic illness, Hunter and colleagues (2011) encouraged participants to keep personal journals as well as take photographs. Sometimes materials are added to address multiple stakeholders' research aims and recognize other epistemologies that may differ from conventional 'western' forms of knowledge. In Young and colleagues' (2013) work in co-creating an 'Aboriginal children's health and well-being measure' with Anishinabek children and families, concepts suggested by the project's advisory committee, community engagement sessions, and an Australian Aboriginal measure of youth mental health were included for discussion along with participants' thematically grouped photographs.

As evidenced above, the use of photovoice with Indigenous communities has seen rapid uptake. Since Castleden and colleagues modified the use of photovoice in 2008 to fit within their community context, variations have been used in Canada to examine the experiences of Indigenous young people with obesity (Pigford, et al., 2013), food security (Ford, et al., 2013; Genuis, et al., 2014) and body image (Shea, et al., 2013), and has been employed to develop culturally appropriate tools for assessing Indigenous children's health and well-being (Young, et al., 2013). The method has been used in countries with similar colonial backgrounds (i.e., Australia and New Zealand) to support urban Indigenous health (Adams, et al., 2012) and extended multigenerational Indigenous families caring for children with asthma (Jones, et al., 2010). Other innovative props have also begun to emerge in qualitative health research; while photovoice is a rather simplistic yet powerful tool, more sophisticated options are now available to us as technology becomes more accessible. Digital stories are an example of such sophistication.

Digital storytelling

Digital storytelling is only just beginning to see uptake in health geography and other health-related research. Digital stories are essentially short (two to five minutes) multimedia film vignettes that draw upon still frame images, movies, audio, music and a pre-recorded narrative often developed through the use of group-led story circles to tell a personal or collective story. Digital storytelling is attributed to Ken Burns' 1990 series of documentaries *The Civil War*. Drawing from the archives of still images, first person narratives, and soundtracks, and combining these onto a single film, Burns demonstrated the affect that could be harnessed in viewers by combining these layers into a single story. Prior to this, and arguably culminating from Paulo Freire's (1971) popular education strategies and philosophies, educators and art practitioners have long attempted to make artistic creation more accessible and purposeful in social change movements. With technology becoming increasingly available to non-experts and a belief that art should not be limited to only the 'gifted' or professionals, the San Francisco Digital Media Center, which would become the current enterprise of StoryCenter, opened its doors in 1994, offering community workshops to teach digital story-making skills (StoryCenter, 2015).

Similar to photovoice, there are a number of steps involved in making a digital story once desire has been expressed to do so by story makers and research partners. The following stages, the first three of which are virtually identical to the process used to initiate a photovoice project, are adapted from the StoryCenter's *Digital Storytelling Cookbook* (Lambert, 2010):

1 Identify a target audience (e.g., policymakers or community leaders) who has the decision-making power that can address situations in question when social change is the end goal of a story making process.
2 Recruit project participants. The researcher or team needs to determine the equipment needs and available support (discussed below) to fit with the ideal number of participants. For example, six to ten people are ideal in terms of in-depth discussion, but equipment and support may reduce the number of participants.
3 Hold a first meeting with participants. Familiarize participants with the project aims and basic ethical principles of digital storytelling (discussed below), explain ethical procedures and informed consent, pose initial themes for developing stories (here the proposed themes are discussed and agreed upon by the participants and the researcher/facilitator), discuss the major elements of digital storytelling (see Figure 10.1) and distribute required equipment (e.g., digital still or video cameras) with instruction on their use.
4 Provide time for participants to develop a written narrative for their story (ideally not more than 500 words) and collect 20–30 pictures, short (5–10 second) video clips, as well as 2–3 musical pieces (ideally without lyrics). Time to compose and complete stories will vary according to story makers' and researchers' schedules.
5 Hold a second meeting with participants to share their stories and for individual story makers to receive feedback from the group (a potential first phase of data collection; seek perspectives from storytellers about the content and process, and identify the key issues, themes or theories from their discussion about their stories).
6 Hold a workshop with participants to voice-record their stories or videotape their narration, create storyboards (see Figure 10.2), import their images, video and music into computer software, and edit and refine the final assembly of their stories to their satisfaction.
7 Screen digital stories within the group and debrief as well as plan a format to share digital stories with target audience(s) (a potential second phase of data collection).
8 Screen digital stories with target audience(s) (a potential third phase of data collection).

There are also a number of key points to consider when using digital story-making and storytelling as a prop in either CBPR or qualitative health research. Our focus below is to illuminate the support needed for a group to go through the story-making process, from script composition to final copy.

The *Seven Elements* have been adapted from the Centre for Digital Storytelling (2010). Watch the four minute video online for more on these points: www.youtube.com/watch?v=NipDAd3_7Do

1 **Point of View**
What is the main point of your story? What is the perspective of the storyteller? Why is it being told now?

- Questions to ask yourself: What realization or point are you trying to communicate with your story? Why it is important for you to tell your story? What do you hope your audience will understand about what you share?

2 **A Dramatic Question**
A key question will guide the story, keep viewer's attention, and will be answered by the end of the story.

- Questions to ask yourself: What was realized by you or by participants in the story? Was there a specific moment or event that was key to your realization? Avoid generalization.

3 **Emotional Content**
Your story may be serious, comical, uplifting, or critical. How will this be presented to attract the audience? How do you want to convey this content?

- Questions to ask yourself: What were you expecting, what happened?

4 **The Gift of Your Voice**
Using your voice helps to personalize the story, but it also contextualizes the content and has a powerful impact.

- Recommendations: Tell your story like you were speaking to a friend, don't read the script. Speak slowly and clearly. Pause when necessary and use emphasis when it is needed.

5 **Soundtrack**
Music or other sounds help to set the tone of the story and add emphasis where needed. Songs can have a greater impact and deep meaning that further personalizes and contextualizes the story.

6 **Keeping it Simple**
Use enough content to tell your story, with pauses used for pictures and scripts as necessary. Your story is conveying a key point, point of view, and/or realization. Be sure not to overload the audience or yourself!

- Recommendations: Refer back to your key point on occasion. Make sure that you are still on track.

7 **Pacing**
Be mindful of the rhythm of the story and how slowly or quickly it progresses. Digital stories are meant to be short: 2-5 minutes.

Figure 10.1 Seven elements of digital storytelling

Digital Storytelling Storyboard

The storyboard is one way to design your digital story. By identifying your pictures in the spaces below, and indicating what text, audio, and effects you would like to correspond with each picture, your story will unfold. It also provides a simple, effective, and adaptable guide for when it comes time to make your story on the computer.

Script:
(Begins with)

(Ends with)

Audio:

Caption:

Transition:

Script:
(Begins with)

(Ends with)

Audio:

Caption:

Transition:

Figure 10.2 Digital storytelling storyboard

Equipment needs

1 One computer per storyteller (individual or group). Free computer software, such as iMovie (Macs; iPads), FinalCut (Mac), MovieMaker (PCs), or PowerPoint (Mac and PCs), can be used to create a digital story. Experience has led us to suggest that ideal conditions would have all participants using the same version of the same software (our preferred version is iMovie on a Macbook).
2 One digital camera (or smart phone) per participant for still images and video.
3 One set of headphones per participant.
4 One high-quality voice recorder.
5 One data projector, portable screen (a white sheet with tacks will do) and good quality speakers.
6 One portable high-quality scanner (optional, if photos are not digital or if participants want to bring in pictures from their personal albums).
7 Access to high-speed internet (optional, needed primarily if images or audio are to be downloaded).
8 Spare batteries for cameras, spare power cords for computers and spare adaptors for all technology.

Workshop environment:

Although the widely used *Digital Storytelling Cookbook* (Lambert, 2010) suggests workshops be designed over a three-day period, they may vary in format, length, availability of technology or story makers interest and level of comfort with specific aspects of the story-making process. In fact, the entirety of the story-making process can be, and arguably must be, adapted to suit the research context. Our experiences have shown that what is required when using digital storytelling in community-based contexts is flexibility, adaptability, and even spontaneity; indeed, Vanessa and Heather, who have both facilitated a number of digital storytelling workshops, have never run two exactly alike.[2] Even the environment that will best house the workshop may change in response to the interests and needs of the group. For instance, when working with youth we have found that holding drop-in times in locations that offered recreational activities (e.g. pool table, foosball table) was ideal. Youth were able to attend when possible, but also simply be in a space with their peers when they were not working on their story.

Story makers will each bring a different set of skills to process. Whereas some may feel more comfortable preparing the script, others may be drawn to gathering pictures or video, compiling music, or perfecting effects for their story. Recalling that digital story-making is a creative and personal process, researchers should be keenly attuned to the strengths that each story maker brings to

2 Classroom settings can provide a more structured environment for digital storytelling. Workshops can be more predictable and scheduled more concretely. See Castleden et al., 2013; Fletcher and Cambre, 2009, for more on the use of digital storytelling in classroom settings.

a digital storytelling workshop; these strengths may complement that of other story makers, thus providing the opportunity for peer-support and mentoring, as each person makes their own story.

Data collection/story composition

The written narrative is a key element of any digital story, but it is often the most difficult creative task. With the brevity of digital stories, it is important to keep in mind that one page of written text will amount to approximately five minutes of narrative once all introductions and transitions are included in the final story. Seven elements have been identified as integral to the digital story-making process (see Figure 10.1). Prior to beginning the script writing process, facilitators might consider sharing these elements, in addition to screening samples of digital stories to help story makers envision what their story could look like in its final edited state.

To support story makers in the creation process, holding focus groups, sharing circles or group discussions around the main theme of the digital stories may be appropriate. This opportunity for open dialogue will allow story makers to hear from their peers, while also gaining feedback on their own ideas for their story. Other strategies to streamline often-diverse thoughts into an identifiable narrative include the use of mind mapping to encourage story makers to envision various aspects of their emerging narrative. Through these generative activities, story makers should determine the point of view or main point that they wish to convey in their stories (Element 1) and the main question that will guide viewers through the story once it is complete (Element 2). The main question, or key point, proposed by the story maker should be emphasized or answered by the end of the story. Once scripts have been compiled and recorded, pictures and soundtracks can be selected to complement the narrative. Although copyright infringements do not usually present problems since digital stories for research are not used for commercial purposes, facilitators should ensure that all credits are provided appropriately.

A storyboard is often useful for story makers to design their story; storyboarding is a process of mechanically planning to plot where each sound bit or effect, picture and portion of script will appear (see Figure 10.2). Story makers pencil in file numbers in each frame, with the beginning and end of each recorded narrative (either by time stamp or words) that will coincide with each picture, in addition to desired transitional effects (i.e., one picture to fade into the next), audio adjustments or possible captions. The storyboard is thus a template to build a digital story; however, it can also be used as a guide to direct the digital construction process. Facilitator involvement in the storyboarding process provides a rich opportunity to engage with story makers about their decision-making processes behind selecting pictures, dividing narratives, or including transitions. Storyboarding can also create an opportunity for editing prior to story composition, as each software program capable of digital storytelling presents its own unique quirks, effective storyboarding can prevent discouragement when editing is attempted while still learning the video editing software. Workshop facilitators are thus strongly encouraged to familiarize themselves with the software that they will be using prior to commencing any workshop.

Analysis and dissemination

Digital storytelling as a research prop requires a story maker's involvement throughout the research process, from design to dissemination. Each of these phases, in addition to the final story, have the potential to generate 'data'. From the content selection to the screening, a digital story involves deep engagement from the story maker, as whatever is included (and excluded) is done at their discretion. Through this creative process, story makers and researchers share an interactive and iterative relationship, whereby the researcher provides technical support, the researcher and other participants contribute to the story-making process and the researcher observes multiple stages of story creation.

Underlying motives feeding into the storyline are often revealed by story makers during this engagement. Field notes and participant observations can aid researchers in documenting this creation process and, later, analysis. Once the stories are complete, researchers may choose to analyze stories for semiotic or discursive elements, both of which have a rich tradition in qualitative research. Due to its novel approach, a unique opportunity for the co-analysis of digital stories, whereby the story maker analyzes the story *for* or *with* the researcher vis-à-vis construction, emerges. Co-analysis additionally mitigates the insertion or potential misinterpretation of story elements by the researcher if analysis were conducted on an individual basis. When disseminating research involving digital stories, researchers have the opportunity to keep story makers voices intact vis-à-vis selecting an online journal that is capable of directing readers to the stories themselves. Time-stamping points of example or including still images of the story in the dissemination of research results further enables the full participation of story makers, even after the dissemination of results. We have employed this strategy after conversation and direction from the story makers and research partners when working within a CBPR context (see Sloan Morgan et al., 2014). Copyright issues continue to be a grey area with respect to using online content but as stated earlier, the general consensus of the academic community is that research participants are producing personal digital stories for educational and not-for-profit purposes and, therefore, with appropriate recognition in the story, free from copyright infringement.

Digital storytelling has emerged from a tradition of 'complex intellectual engagement that is at once creative, socially oriented, and pedagogical' (Fletcher and Cambre, 2009, p. 111). As a tool in education, digital storytelling is being used to promote deep learning, whereby story makers enter a process of reflexivity and situatedness (e.g. Castleden, et al., 2013). But as a research method, participant-led film-making is still awkwardly situated in the social sciences (Jacobs, 2013). Yet we are beginning to see this approach, in the form of digital storytelling, as a methodological prop to decenter the research process while ensuring that community voices, perspectives, and representations remain at the core of inquiry (e.g. Cunsolo Willox, et al., 2013; Sloan Morgan, et al., 2014). Within geography, where we are seeing digital storytelling most frequently is in the area of Indigenous research (e.g. Iseke and Moore, 2011); Offen (2012) refers to this as 'Indigital Geographies'. In addition, programs working with populations who have experienced trauma have used digital storytelling to promote healing through sharing difficult experiences or by vulnerable populations re-creating events that impeded health and well-being (e.g. Prairie Women's Health Centre of Excellence, 2011; Silence Speaks, 2015; Stout and Peters, 2011).

Interestingly, health geographers have yet to seriously engage with the potential limitations or ethical tensions associated with digital storytelling as a method for collecting data, both in the field of health geography and related disciplines. But health researchers, specifically Gubrium and colleagues (2014), discuss six key challenges. First, ambiguous boundaries exist between research, practice, and advocacy, which are often entangled in non-conventional and participatory approaches to research. Second, when recruitment and consent to participate in research on often stigmatizing or traumatic health issues or involving marginalized populations risks further ostracization, care must be taken to not perpetuate further harm. Third, workshop facilitators (often the researcher or research team) may unduly influence participants' story creations, shaping them into stories that may resonate with wide (including academic) audiences but nevertheless risk the loss of participants' voices. Researchers need to pay constant attention to the power dynamics associated with the process. Fourth, stories often include more than just the story maker: how others are represented and the potential harm that may be inflicted are important considerations that may be interpreted by the participant to be censorship. Fifth, confidentiality is not something that can be easily achieved in the digital story-making process; this necessitates exploring the tensions between personal privacy and ability to maintain such knowledge production as participants' own intellectual property. Sixth and finally, the release of digital stories for use in multiple academic and public venues requires an ongoing consent process throughout the project and beyond to the dissemination.[3] Despite these tensions, digital storytelling has the potential to incorporate spatiality, memory and reflexivity, all aspects of health research in geography that can deepen understandings of health as a lived experience embedded in situated contexts.

3 Gubrium and colleagues (2014) point to several other important contributors on this topic, while not specific to health geography, certainly aid in our scholarly engagement with the prop: Cunsolo Willox, A., Harper, S. L., and Edge, V. L., (2013), 'Storytelling in a digital age: digital storytelling as an emerging narrative method for preserving and promoting indigenous oral wisdom', *Qualitative Research*, 13(2), pp. 127–147; Dush, L. (2013), 'The ethical complexities of sponsored digital storytelling', *International Journal of Cultural Studies*, 16(6), pp. 627–640; Pittaway, E., Bartolomei, L., and Hugman. R., (2010), '"Stop stealing our stories": the ethics of research with vulnerable groups', *Journal of Human Rights Practice*, 2(2), pp. 229–251; and Hill, A. L., (2010), 'Digital storytelling for gender justice: exploring the challenges of participation and the limits of polyvocality', in Bergoffen, D., Gilbert, P. R., Harvey, T., and McNeely, C. L. (eds.), *Confronting Global Gender Justice*. Oxford, UK: Routledge, pp. 126–140.

Reflections on praxis (part 2): five common themes

Having used and reflected on photovoice and digital storytelling as props ourselves over the past several years in health-focused CBPR, we have identified five common themes about their utility: (1) implementing partner preferences benefits the research and the research relationship, (2) analysis of audiovisual data is often overlooked, (3) audiences for research outputs using these props are often more far-reaching than conventional methods, (4) a number of ethical tensions and institutional hurdles associated with the use of these props must be considered and navigated with care, and (5) technological considerations are key to the effective use of these props. Each of these is discussed briefly below.

In CBPR, partner preferences matter

Community-based partners bring a wealth of knowledge about their communities, including political and interpersonal dynamics, and preferences for knowledge generation and mobilization. All of these are invaluable in terms of guidance throughout every stage of the research project, from the formation of research questions to dissemination processes targeting appropriate audiences (see Nykiforuk, et al., 2011). Establishing a local project advisory committee adds additional community-level control over the research process and brings important insights into how data are to be analyzed, impacting the data set and creating an analytical process more clearly linked to achieving or altering stated community aims (Young, et al., 2013). At the same time, an Advisory Committee – like the researcher or team – also holds potential sway over participants' autonomy and personal goals for participation in unintended ways, thus reflexive engagement with this potentiality is critical throughout the research cycle. Finally, while interviews or other conventional qualitative research methods may seem like the logical way forward from an academic point of view, innovation through props like the ones described above may be the difference between a community partnership moving forward, or not. In short, utilizing props over conventional research methods, such as interviews, has the potential to provide not only greater insight but also, and perhaps more importantly, contributes to community members' desire for autonomy to decide *how* they would like to answer the questions they have determined will contribute towards researching themselves 'back to life'.

Analysis of the audiovisual data should not be overlooked

As props, photovoice and digital stories are often the tools to elicit similar data from participants as if they were being interviewed or participating in a focus group. Transcriptions of interviews inspired by the creation process then become the principal data set subjected to analysis, but the visual and audio data that are produced are also ripe for analysis. We do not, however, see this audiovisual data being analyzed very often in the literature, so we draw your attention to some exemplars. In a photovoice-based study of sanitation, water, and health in rural

Kenya, a research team member familiar with the community examined participant photographs. Recurrence of key notable features was then used as a proxy of importance for that attribute, for example, the presence of water storage containers, firewood or animals in the photograph (Levison, et al., 2012, p. 191). This study also coded participants' hand-drawn maps. While 'art' in various forms has been the subject of analysis in qualitative research for a very long time, there is little procedural uniformity when it comes to coding such audiovisual data. However, qualitative data analysis software is now more and more able to accommodate electronic audiovisual material. For example, Bisung and colleagues (2015) coded their participant photographs directly using qualitative data analysis software (in this case, NVivo 10), then generated more themes through coding participant interviews.

Audiovisual elements increase audiences for research outputs

Effective knowledge translation is critical to the success of any research project; but academics have traditionally focused on peer-reviewed journal articles and conference papers to convey this new knowledge. The utility of the props discussed in this chapter goes beyond data elicitation – they can be used (and, in fact, were originally designed) to effectively communicate across a broad range of audiences by sharing the photography and stories themselves along with the writings generated from their analysis. Presenting some combination of 'raw' material and interpreted outcomes, determined in agreement with research participants, can be an effective way of encouraging dialogue with multisectoral stakeholders. Visual materials are often presented where both research participants and those with decision-making power might mingle, including educational and health care facilities. For example, after a research team employed photovoice with a group of Indigenous youth on the issue of food security in Alberta, Canada, they circulated a photobook around the local school and public health fair (Genuis, et al., 2014). Another project, which engaged African Americans with chronic illnesses, created a touring exhibition for regional hospitals and health centers (Hunter, et al., 2011). Our own digital storytelling research with Indigenous youth in British Columbia, Canada, for example, led to the submission of their stories to the annual, international imagineNATIVE film festival and directly informed community-level decision-making and land-use planning (these stories can be viewed at http://www.heclab.com/digitalstories/huu-ay-aht-youth-digital-stories/).

Ethical considerations

In our experience, as qualitative research props, digital storytelling and photovoice involve relational ethics, not just procedural considerations specific to *both* CBPR *and* visual documentary methods. This includes the imperative to build trusting, equitable research relationships, and this usually needs to happen over an extended period of time (Castleden, et al., 2012). As Nykiforuk and colleagues (2011) note, a lack of relationship building prior to beginning the process can create ethical dilemmas about raising awareness and expectation among participants and other community partners. These props also introduce the ethical challenge of 'unique relationships between the researcher, the research participants (photographers) and those photographed' (Bisung, et al., 2015, p. 210). Although there exists a growing literature on the ethical practice of CBPR, visual methods and qualitative methods generally (e.g. Flicker, et al., 2007; Castleden, et al., 2008; Gubrium, et al., 2014), a review of studies employing these props highlights specific concerns, including storing images depicting illegal activity (Dennis, et al., 2009), the personal safety of mobility-impaired participants (Newman, et al., 2010), the potential emotional distress for participants revisiting painful stories and events (Hunter, et al., 2011), and the ethical decision to publish (or not) participants' visual data, rather than only aggregate 'findings' (Clark, et al., 2010; Levison, et al., 2012). Issues associated with relational and procedural ethical considerations in research, particularly CBPR, are an emergent aspect of the research enterprise. With the use of props such as those we have described here, additional interrogation is required in order to ensure the core principles of research with integrity are upheld while the autonomy of research participants is also respected.

Technological considerations

While much of the technology required for working with photovoice and digital storytelling is becoming more affordable and accessible, there are still many technical decisions to consider. Many photovoice studies have used disposable film cameras, but the popularity of smartphones, depending on the economic context, is quickly rendering the 'digital versus 35mm film camera' debate a thing of the past. The benefit (and limitation) of disposable cameras for photovoice is that pictures cannot be erased and there is a limit to how many can be taken. This creates opportunity to explore why a picture was initially considered, but then deemed not as important as others. It also encourages critical thinking on the part of the photographer/participant to 'not use up' their exposures. At the same time, this interferes with their privacy in the instances where their preference was to delete a particular photograph, but they are unable to do so with a disposable camera. They may also experience frustration with a 'wasted' picture when they are limited with how many they can take.

Some basic principles for making digital storytelling enjoyable, not only for participants but also for those facilitating the process include making sure everyone who needs it has the same computer software, and that people are comfortable in whichever computer platform (PC or Mac) is used. Depending on the number of participants and type of equipment used in either photovoice or digital storytelling, you may need to make extra provision for technical training and support. Finally, as with any research method relying on recording technology, there are perils. In our experience, for example, audio recordings can, at times, be frustratingly poor in sound volume and quality.

Concluding comments

Of course, the key question in regards to research with communities using arts-based methods or conventional methods (i.e., interviews) remains: 'To what end(s)?' We have found that the use of these props, and arts-based participatory methods generally, will lead to not only the creation of new knowledge but also new questions and exciting new theoretical, methodological, substantive, and ethics-oriented research trajectories. More importantly though, we have found that the communities we work with have embraced these props for their own purposes soon after our research projects have ended. For the potential praxis of participatory arts-based research to be realized, these new questions need to be not only expected but also embraced. If not, then we are at risk of being left with only an 'information-eliciting' aim to arts-based approaches rather than the rich, multi-sensory experience that they are theoretically intended to provide. In our varied experiences, arts-based participatory methods enable multiple entry points to collectively approaching research questions, not just through, or for, eliciting data. In closing, we remind readers to embrace research innovation and 'play' with what we have outlined above about how we have used these qualitative research props. In doing so, you will undoubtedly take what has been done in new directions.

References

Adams, K., Burns, C., Liebzeit, A., Ryschka, J., Thorpe, S., and Browne, J., 2012. Use of participatory research and photo-voice to support urban Aboriginal healthy eating. *Health and Social Care in the Community*, 20, pp. 497–505.

Ball, J., and Janyst, P., 2008. Enacting research ethics in partnerships with Indigenous communities in Canada: 'Do it in a good way'. *Journal of Empirical Research on Human Research Ethics,* 3(2), pp. 33–51.

Bisung, E., Elliott, S. J., Abudho, B., Schuster-Wallace, C. J., and Karanja, D., 2015. Dreaming of toilets: using photovoice to explore knowledge, attitudes and practices around water-health linkages in rural Kenya. *Health and Place*, 31, pp. 208–215.

Brant Castellano, M., 2014. Ethics of Aboriginal Research. In Teays, W., Gordon, J., and Dundes Renteln, A. (Eds). *Global Bioethics and Human Rights: Contemporary Issues.* Lanham, MA: Rowman and Littlefield. pp. 273–288.

Brant Castellano, M., and Reading, J., 2010. Policy writing as dialogue: drafting an Aboriginal chapter for Canada's Tri-Council Policy Statement: ethical conduct for research involving humans. *The International Indigenous Policy Journal*, 1(2), p. 1.

Burgess, J., 2006. Hearing ordinary voices: cultural studies, vernacular creativity and digital storytelling. *Continuum: Journal of Media and Cultural Studies*, 20(2), pp. 201–214.

Cameron, E., 2012. New geographies of story and storytelling. *Progress in Human Geography*, 36(5), pp. 573–592.

Carroll, W., 2004. Inquiry as empowerment: participatory action research. In W. K. Carroll (ed.), *Critical Strategies for Social Research.* Toronto, Ontario: Canadian Scholar's Press. pp. 276–280.

Castleden, H., Garvin, T., and Huuy-ay-aht First Nation, 2008. Modifying Photovoice for community-based participatory Indigenous research. *Social Science and Medicine*, 66, pp. 1393–1405.

Castleden, H., Sloan Morgan, V., and Lamb, C., 2012. 'I spent the first year drinking tea': Exploring Canadian university researchers' perspectives on community-based participatory research involving Indigenous peoples. *The Canadian Geographer / Le Géographe Canadien*, 56(2), pp. 160–179.

Castleden, H., Daley, K., Sloan Morgan, V., and Sylvestre, P., 2013. Settlers unsettled: using field schools and digital stories to transform geographies of ignorance about Indigenous peoples in Canada. *Journal of Geography in Higher Education*, pp. 1–13.

Clark, A., Prosser, J., and Wiles, R., 2010. Ethical issues in image-based research. *Arts and Health: An International Journal for Research, Policy and Practice*, 2(1), pp. 81–93.

Cunsolo Willox, A., Harper, S. L., Edge, V. L., Landman, K., Houle, K., and Ford, J. D., 2013. The land enriches the soul: On climatic and environmental change, affect, and emotional health and well-being in Rigolet, Nunatsiavut, Canada. *Emotion, Space and Society*, 6, 14–24.

de Leeuw, S., Cameron, E. S., and Greenwood, M. L., 2012. Participatory and community-based research, Indigenous geographies, and the spaces of friendship: a critical engagement. *The Canadian Geographer / Le Géographe canadien*, 56(2), 180–194.

Dennis S., Jr., Gaulocher, S., Carpiano, R. M., and Brown, D., 2009. Participatory photo mapping (PPM): exploring an integrated method for health and place research with young people. *Health and Place*, 15, 466–473.

Fals-Borda, O., and Rahman, A., 1991. *Action and knowledge: breaking the monopoly with participatory research.* New York: The Apex Press.

Fletcher, C., and Cambre, C., 2009. Digital storytelling and implicated scholarship in the classroom. *Journal of Canadian Studies*, 43(1), pp. 109–130.

Flicker, S., Travers, R., Guta, A., McDonald, S., and Meagher, A., 2007. Ethical dilemmas in community-based participatory research: Recommendations for institutional review boards. *Journal of Urban Health*, 84(4), pp. 478–493.

Ford, J. D., Lardeau, M.-P., Blackett, H., Chatwood, S. and Kurszewski, D., 2013. Community food program use in Inuvik, Northwest Territories. *BMC Public Health* 2013, 13, 970.

Fox, N., 2003. Practice-based evidence: towards collaborative and transgressive research. *Sociology*, 37(10), pp. 81–102.

Freire, P., 1971. *Pedagogy of the oppressed.* New York, NY: Continuum Press.

Genuis, S. K., Willows, N., Alexander First Nation, and Jardine, C., 2014. Through the lens of our cameras: children's lived experience with food security in a Canadian Indigenous community. *Child: Care, Health and Development*, 1–11.

Gubrium, A. C., Hill, A. L., and Flicker, S., 2014. A situated practice of ethics for participatory visual and digital methods in public health research and practice: a focus on digital storytelling. *American Journal of Public Health*, 104(9), pp. 1606–1614.

Hesse-Biber, S., and Yaiser, M., 2004. *Feminist perspectives on social research.* New York: Oxford University Press.

Hunter, J., Langdon, S., Caesar, D., Rhodes, S. D. and Pinkola Estés, C., 2011. Voices of African American health: stories of health and healing. *Arts and Health: An International Journal for Research, Policy and Practice*, 3(1), pp. 84–93.

Iseke J., and Moore, S., 2011. Community-based indigenous digital storytelling with elders and youth. *American Indian Culture and Research Journal*, 35, pp. 19–38.

Jacobs, J., 2013. Listen with your eyes: towards a filmic geography. *Geography Compass*, 7, pp. 714–728.

Jacklin, K., and Kinoshameg, P., 2008. Developing a participatory aboriginal health research project: 'Only if it's going to mean something'. *Journal of Empirical Research on Human Research Ethics*, 3(2), pp. 53–67.

Jones, B., Ingham, T., Davies, C. and Cram, F., 2010. Whānau Tuatahi: Māori community partnership research using a Kaupapa Māori methodology. *MAI Review*, 3, pp. 1–14.

Kauper-Brown, J., and Seifer, S., 2006. *Developing and sustaining community-based participatory research partnerships: a skill building curriculum*. University of Washington: The Examining Community-Institutional Partnerships for Prevention Research Group.

Keeshig-Tobias, L., 1997. Stop stealing Native stories. In Ziff, B., and Rao, P. (eds.), *Borrowed power: essays on cultural appropriation*. New Brunswick, NJ: Rutgers University Press. pp. 71–73.

Kirby, S., and McKenna K., 1989. *Experience, research, social change: methods from the margins*. Toronto, ON: Garamond Press.

Kindon, S., Pain, R., and Kesby, M., 2007. Participatory action research: origins, approaches, methods. In S. Kindon, R. Pain, and M. Kesby (ed.), *Participatory action research approaches and methods: connecting people, participation and place*. New York, NY: Routledge. pp. 9–18

Koster, R., Baccar, K., and Lemelin, R. H., 2012. Moving from research ON, to research WITH and FOR Indigenous communities: A critical reflection on community-based participatory research. *The Canadian Geographer / Le Géographe canadien*, 56, pp. 195–210.

LaFrance, J., and Crazy Bull, C., 2009. Researching ourselves back to life: taking control of the research agenda in Indian country. In Mertens, D., and Ginsberg, P. (eds.). *The Handbook of Social Research Ethics*. Thousand Oaks, CA: Sage Publications. p. 135–150.

Lambert, J., 2010. *Digital storytelling cookbook*. Berkley, CA: Digital Diner Press. Available at: http://storycenter.org/cookbook-download/.

Lambert, J., 2013. *Digital storytelling: capturing lives, creating community*. 4th ed. New York, NY: Routledge.

Levison, M., Elliott, S., Schuster-Wallace, C., and Karanja, D., 2012. Using mixed methods to visualize the water-health nexus: identifying problems, searching for solutions. *African Geographical Review*, 31(2), pp. 183–199.

Lewin, K., 1946. Action research and minority problems. *Journal of Social Issues*. 2, pp. 34–46.

Mejia, A. P., Quiroz, O., Morales, Y., Ponce, R., Limon Chavez, G., and Olivera y Torre, E., 2013. From madres to mujeristas: Latinas making change with Photovoice. *Action Research*. 11(4), pp. 301–321.

Minkler, M., and Wallerstein, N., 2003. *Community-based participatory research for health*. San Francisco: John Wiley.

Mitchell, E. M., Steeves, R., and Dillingham, R., 2014. Cruise Ships and Bush Medicine: Globalization on the Atlantic Coast of Nicaragua and Effects on the Health of Creole Women. *Public Health Nursing*, 32(3), pp. 237–245.

Mitchell, E. M., Steeves, R., and Perez, K. H., 2015. Exploring Creole women's health using ethnography and Photovoice in Bluefields, Nicaragua. *Global Health Promotion*, 22(4), pp. 29–38.

Newman, S., 2010. Evidence-based advocacy: using photovoice to identify barriers and facilitators to community participation after spinal cord injury. *Rehabilitation Nursing*, 35(2), pp. 47–59.

Nykiforuk, C., Valliantos, H., and Nieuwendyk, L., 2011. Photovoice as a method for revealing community perceptions of the built and social environment. *International Journal of Qualitative Methods*, 10(2), pp. 103–124.

Offen, K., 2012. Historical geography II: digital imaginations. *Progress in Human Geography*, 37(4), pp. 564–577.

Pigford, A. A., Ball, G. D. C, Plotnikoff, R. C., Arcand, E., Alexander First Nation, Fehderau, D. D., Holt, N. L., Veugelers, P. J., and Willows, N., 2013. Community-based participatory research to address childhood obesity: experiences from Alexander First Nation in Canada. *Pimatisiwin: A Journal of Aboriginal and Indigenous Community Health*, 11(2), pp. 171–185.

Pittaway, E., Bartolomei, L., and Hugman, R., 2010. 'Stop stealing our stories': the ethics of research with vulnerable groups. *Journal of Human Rights Practice*, 2(2), pp. 229–251

Prairie Women's Health Centre of Excellence, 2009. Digital stories – First Nations women explore the legacy of residential schools. Available at: http://www.pwhce.ca/program_aboriginal_digitalStories.htm.

Price, P. L., 2010. Cultural geography and the stories we tell ourselves. *Cultural Geographies*, 17(2), pp. 203–210.

Razack, S., 1993. Story-telling for social change. *Gender and Education*, 5(1), pp. 55–70.

Rose, G., 1997. Engendering the slum: photography in East London in the 1930s. *Gender, Place and Culture*. 4(3), pp. 277–301.

Rule, L., 2010. Digital storytelling: never has storytelling been so easy or so powerful. *Knowledge Quest*, 38(4), pp. 56–57.

Schnarch, B., 2004. Ownership, control, access, and possession (OCAP) or self-determination applied to research. *Journal of Aboriginal Health*, 1(1), pp. 80–95.

Shea, J., Poudrier, J., Thomas, R., Jeffrey, B., and Kiskotagan, L., 2013. Reflections from a creative community-based participatory research project exploring health and body image with First Nations girls. *International Journal of Qualitative Methods*, 12, pp. 272–293.

Sherman, M., Berrang-Ford, L., Ford, J., Lardeau, M., Hofmeijer, I., and Zavaleta Cortijo, C., 2012. Balancing indigenous principles and institutional research guidelines for informed consent: a case study from the Peruvian Amazon. *AJOB Primary Research*, 3(4), pp. 53–68.

Silence Speaks, 2015. About. Available at: http://silencespeaks.org/.

Sloan Morgan, V., Castleden, H., and Huu-ay-aht First Nations, 2014. Redefining the cultural landscape in British Columbia: Huu-ay-aht youth visions for a post-treaty era in Nuu-chah-nulth territory. *ACME: An International E-Journal for Critical Geographies*, 13(3), pp. 551–580.

StoryCenter, 2015. How it all began. Available at: http://storycenter.org/history/.

Stout, R., and Peters, S., 2011. *kiskinohamâtôtâpânâsk: Inter-generational effects on professional First Nations women whose mothers are Residential School Survivors* (Project Summary). Prairie Women's Health. Available at: http://www.pwhce.ca/pdf/kiskino.pdf.

Sylvester, R., and Greenidge, W., 2012. Digital storytelling: extending the potential for struggling writers. *The Reading Teacher*, 63(4), pp. 284–295.

Tandon, R., 2002. *Participatory research: revisiting the roots*. New Delhi: Mosaic Books.

Teelucksingh, C., and Masuda, J. R., 2014. Urban environmental justice through the camera: understanding the politics of space and the right to the city. *Local Environment: The International Journal of Justice and Sustainability*, 19(3), pp. 300–317.

Valiquette-Tessier, S.-C., Vandette, M.-P., and Gosselin, J., 2015. In her own eyes: photovoice as an innovative methodology to reach disadvantaged single mothers. *Canadian Journal of Community Mental Health*, 34(1), pp. 1–16.

Wang, C., and Burris, M.A., 1997. Photovoice: concept, methodology, and use for participatory needs assessment. *Health Education and Behavior*, 24(3), pp. 369–387.

Wang, C., 1999. Photovoice: a participatory action research strategy applied to women's health. *Journal of Women's Health.* 8(2), 185–192.

Wang, C., and Redwood-Jones, Y., 2001. Photovoice ethics: perspectives from Flint Photovoice. *Health Education and Behavior*, 28(5), pp. 560–572.

Young, N., Wabano, M. J., Burke, T. A., Ritchie, S. D., Mishibinijima, D., and Corbiere, R. G., 2013. a process for creating the Aboriginal children's health and well-being measure (ACHWM). *Can J Public Health*, 104(2), pp. 136–141.

11 Media and framing

Processes and challenges

S. Michelle Driedger and Theresa Garvin

Reflections on praxis

At first glance, conducting media analyses in health geography may seem simple: search out sources, bring the data together, analyze it and report findings. However, we have found that each of these steps is fraught with challenges for both traditional and new (e.g., online) media sources. Researchers must make choices about which sources to include and, more importantly, which ones to exclude, how to analyze the data in a rigorous manner and then determine the most effective way to communicate results. Ethical considerations run throughout.

For print media, for example, this means asking whether inclusion is as straightforward as automatically including a keyword or set of keywords. However, such broad parameters might result in thousands of pages of text, depending on the topic under study. So then the researcher must ask if two, three or four particular keywords are required or if the keywords must occur within a particular context (i.e., within the same paragraph). Such decisions are not merely methodological; rather, they involve ethical questions including which keywords encapsulate meaning and, importantly, which ones do not. For example, in a study of print media comparing suburban development in Canada versus Ireland, terminology matters. In Canada such developments are called 'suburbs', while in Ireland they are called 'housing estates'. Choice of keywords can bias findings.

Traditional media analyses are facilitated by the ability to search multiple media sources (e.g., newspaper, popular magazines) through available academic e-resource databases. However, such sources themselves can introduce challenges and biases. For example, as part of a standard quality check, Driedger and Weimer (2015) discovered that two commonly available e-resource databases produced different retrieval yields despite using the exact same search parameters. The authors were particularly concerned in their discovery that the *same* e-resource database produced very different retrieval results when accessed through different Canadian university libraries. The reason was eventually located, but it signalled a much larger problem for the reliability of simply collecting a sample data set to analyze. Because Dow Jones's Factiva e-resource database was never designed to serve academic clients, search results can vary dramatically from one university to another, and thus results and conclusions can also vary. Hence, Driedger and Weimer (2015) conclude that for the collection of traditional newspaper media stories, an academic researcher is better served using Proquest's Canadian Newsstand Major Dailies database because it functions more consistently across universities. The lesson learned is that we must be aware of potential biases even in university-available e-resources.

Moreover, in conducting any kind of media analysis, it is important to be attuned to the sources used in the collection of a data set. Each media outlet – whether traditionally based, such as a newspaper, compared to alternative sources for new media, like a blog or a newsfeed – caters to the ideological preferences of its readers. In other words, readers gravitate to media source types that 'fit' their cultural or sociopolitical views of the world. Hence, considerable attention should be provided to the types of media outlets examined, as well as providing reasonable justification for their selection.

For new media, especially blogs or group sources (Facebook, Twitter, Reddit), inclusion and exclusion criteria can be even more critical. For example, a researcher may be challenged on how to handle a Twitter feed that includes hashtags to ideas or feeds peripheral to the research question. Likewise, a Reddit feed can digress into tangential stories not directly related to the concept under investigation. A researcher must enter this research space with a clear intent on what is important: the central story or the tangents? Both might produce important and interesting results, but each answers a different research question. Therefore the research question should focus on a discreet set of keywords, date ranges, clearly defined sources and often geographical considerations (where the source is located or where it is published).

When conducting qualitative media analysis, constant critical reflection must take place to ensure control over researcher bias. It should be noted, however, that careful attention must also be paid to the differences in traditional and new media in the analysis stage. Practice standards in traditional media encourage the inclusion of multiple viewpoints on a particular issue. In many cases, traditional media provides alternate viewpoints to ensure that 'both sides of the story' are represented. By comparison, new media sources may present only one, very limited, perspective. For open forums like Twitter or Reddit, there is limited ability to 'fact-check' what is discussed, yet non-media personnel can actively shape the nature, tone, and content of discussions. Health promotion advocates have been increasingly utilizing platforms such as Facebook to analyze how messages are being interpreted around issues such as breast cancer (Abramson, et al., 2015) and smoking cessation (Rama, et al., 2015). The situated content on these specialized sites provides important understandings of how health messages are interpreted and reinforced.

Moreover, new media sources also introduce different ethical responsibilities than traditional media outlets. Traditional media is considered publicly accessible, and therefore typically does not require research ethics board (REB) approval for study and analysis. By contrast there are issues of sensitivity that need to be attended to for new media. While discussed in greater detail in the ethics section of this chapter, researchers should be mindful of national and institutional guidelines. Importantly, guidelines for the ethical conduct of research change over time as does how individual institutional REBs operationalize and interpret these guidelines. For example, Canada's Tri-Council Policy Statement 2 (TCPS 2) which is the latest edition of ethical conduct requirements for research involving humans, was released December 2014 (CIHR, et al., 2014) and includes new guidelines acknowledging the challenges of new media sources.

Whether based on traditional media or new media, these types of analyses are broadly considered *observational designs*. Observation research is used to study behavior 'in living, natural and complex communities or settings, in physical environments, or in virtual settings' (CIHR, et al., 2014, p.145). These environments can be publically accessible spaces (anywhere an individual can openly access even though there may be certain restrictions in terms of allowable times, the payment of a fee [or not], like a library, park, stadium, etc.), virtual spaces (the Internet, including new media platforms) or private or controlled spaces (places that have controlled access often established through membership). Observational research can be either non-participatory (the researcher just observes as in naturalistic inquiry) or participatory (the researcher actively engages in the scene or setting being observed, and where such participation is fundamentally important for effective analysis).

Where ethical codes of conduct and the formalized requirement for approval by a certified REB for observational research designs may vary, there is one clear level of guidance: if the observation is occurring in a place that has a reasonable expectation of privacy (e.g., religious services, Internet chat rooms) REB approval is required, although there may be exemptions to the requirement to obtain informed consent provided that appropriate privacy protections are in place. It is important to underscore here, that the TCPS 2 policy operationalizes a reasonable expectation of privacy more broadly than might be initially thought by researchers. For example, researchers might rightly identify that REB approval is needed if membership approval is required to join or view a particular online group in order to have content made visible. However, researchers also need to be careful in situations where content is public by default, because privacy controls vary from platform to platform (Batrinca and Treleaven, 2015). Much of our own research has entailed media analyses in conjunction with additional methods, such as interviews and focus groups, and ethics approval has been easily acquired for including traditional and online media sources as part of the larger projects. Unlike other experiences related in this volume, we have found ethics boards to be quite reasonable in their application of regulations in the context of traditional and new media sources. That said, it is always important to liaise with your own ethics board as concerns over what is public and what is private (especially in new media) can directly impact research methods and processes.

Introduction

Traditional[1] and new media are concurrently reflections of public discourse and agenda-setters. Traditional media reflects public discourse by reporting on issues

1 When we use the term 'traditional media' we are referring print, radio and television.

that are part of our collective public discussions; it sets agendas by drawing public attention to issues and by telling the public what they should be paying attention to, while at the same time trying to meet market demand for information and satisfy advertisers' desires for exposure. This idea of what we pay attention to and are guided towards at any given time also applies to new media, but with a twist. Unlike traditional media, new media provides many more opportunities to actively shape, challenge and orient public discourse while at the same time restricting what we even see or think about via hidden filters built into many of the platforms on which new media are based.

The benefit of media analyses is that they offer the opportunity to gain insight into how an issue under study is being presented within civil society discourse. The advantages include (1) the ability to interrogate how multiple media sources might be portraying aspects of the same topic in different ways; (2) the opportunity to examine how different groups within civil society are contesting how issues are being defined and what aspects are deemed most salient; (3) the potential to identify key silences within this discourse, either by virtue of not being given sufficient space or coverage (if it is only marginally represented), or that these issues on which the media are silent have yet to be brought into this medium (if it is not present at all); and (4) a mechanism to trace salience to given topics over time by virtue of how much they are given prominence in media stories, feeds and discussions. This chapter outlines the key activities involved in the conduct of media analyses and discusses how media analyses can provide important insights for health geographers. The chapter will include five sections. The first section provides a background on principles behind a framing analysis of media. The second identifies and explains the key stages for collecting data, including how to access data sources. The third section includes examples of how media data sets can be analyzed and how to effectively report findings. The fourth reviews some ethical challenges in analyzing both traditional media (print, television, radio) and new media (Facebook, Twitter, Reddit). Throughout these sections we offer critical reflections relevant to health geography researchers. Finally, the last section of the chapter will provide some conclusions on media analysis and framing.

Media and the importance of framing

The power of the media in civil society discourse has been long recognized within the literature. Lippmann (1922) was one of the earliest scholars to identify that how members of society make sense of any issue, be it political, global, risk, technological and so forth, is typically formed second-hand through media presentations of a topic as opposed to relying on direct observation and experience. Later, McCombs and Shaw (1972) introduced a theory by which to examine the agenda-setting function of the media, namely, that the media gives salience to particular topics by virtue of what it covers, and to a lesser extent 'how' a topic is covered. The more attention that is given to a topic increases its salience in civil society and has the potential to dramatically shift political agendas. This theory emerged from an examination of the 1968 US Presidential election and has remained prominent (McCoombs and Shaw, 1993; Rogers and Dearing 1988; Soroka 2002). Some scholars have challenged the agenda-setting function of the media as being too one-sided (Lang and Lang, 1981; Driedger, 2008). These critics suggest that the media is only one key actor (albeit a prominent one) alongside the activities of other notable groups (political actors, business interests, grassroots organizations, interest groups) that can increase or diminish the attention given to specific topics. Acknowledging Cohen (1963), that the media might not tell a public what to think, it nonetheless suggests to a public what to think about. Hence, it is important for scholars to move the emphasis of analysis away from pure content to focus more on salience informed by framing theory.

Framing theory underscores the importance of *how* issues are presented in a media text. How media stories are written and the language that is used help to construct and reinforce dominant ideologies designed to draw attention to some aspects of an issue and to detract attention from others (Fleras, 2003; Foucault, 1986). This process is done as 'frame-building' (Sheufele, 1999), such that a frame serves to centrally organize (Gamson, 1989) the content of media stories (Tankard, et al., 1991, as cited in Durfee, 2006). Entman's (1993) definition of framing is frequently used in the literature: 'To frame is to select some aspects of a perceived reality and make them more salient in a communicating text, in such a way as to promote a particular problem definition, causal interpretation, moral evaluation, or treatment recommendation'. Pivotal in such analyses are the framing effects that are employed to draw attention to issues, events or people in particular ways (Iyengar, 1991; Price and Tewksbury, 1997; Sheufele and Tewksbury, 2007). Frames establish boundaries around an issue (Driedger and Eyles, 2003) often to make an issue appear less complex (Kim and Anne Willis, 2007). For example, in examining Canadian media presentations of chlorinated disinfection byproducts in drinking water that are formed by the chlorination process, Driedger and Eyles (2003) concluded that traditional news media framed these stories in two major ways, with the former being more dominant: chlorine is killing us by causing cancer (death frames) versus chlorine saves lives by removing microbial risks in the water supply (health frames). Likewise, health promotion stories are often framed more in terms of 'gains' (from doing the recommended action) over any associated 'losses' (additional important aspects that are equally relevant). Muzyka and colleagues (2012) identified in a prominent Kenyan newspaper that the 'gain' frame of voluntary medical male circumcision in protecting heterosexual men from developing HIV/AIDS did not sufficiently highlight the 'loss' frames that circumcision only offers partial protection and that condom use is still needed.

Collecting media data sets

Collecting media data can be remarkably easy. Collecting media data that will be meaningful to answer a study's research questions and objectives takes careful thought. Since at least the year 2000 it became more commonplace that traditional media data could be accessed through electronic full text databases as opposed to relying on paper-based indexes (e.g., Canadian Business and Current Affairs Index, Canadian News Index, Canadian Periodical Index) with subsequent searching for full-text stories using microfilm and microfiche. Electronic searching of media data with full-text access to content has greatly facilitated the research opportunities available to researchers.

Regardless of the type of media data – traditional or new media – all data sets have inherent limitations that create methodological challenges in collecting data, as well as interpretive challenges in efforts to describe what the collected data means with respect to the study objectives or purpose. The main limitations to be discussed in this section relate to how the data are accessed. For traditional media sources, the data collection limitations are largely dependent on the databases used. For new media data, the data collection limitations relate primarily to comprehensiveness, with interrelated concerns associated with personalization and cost. As in all research, an additional limitation is the capacity of the researcher to conduct the search – in terms of their skills, their creativity and their problem-solving and critical thinking strategies. While this section will provide some guidance for good data collection strategies, researcher capacity comes mainly from experience and experimentation.

Depending on type of media data being accessed, it is important to determine what type of database will help retrieve useful data for analysis. University based libraries have a number of e-resource databases available. Consulting the relevant librarians for suggestions about which may be most appropriate to the study needs is advantageous. For Canadian newspapers, ProQuest's Canadian Newsstand Major Dailies offers comprehensive electronic coverage of national and leading regional newspapers in Canada, although the dates of coverage vary.

By contrast, Dow Jones's Factiva, mostly available at universities with a large School of Business, searches a much wider range of media data – from blogs (of companies, corporations, media outlets), news digests (from various countries and in different languages), several different newspapers from all over the world, newswires and much, much more. When using Dow Jones's Factiva, it is important to be aware of how the database functions and how individual users can set search parameters that affect all subsequent searches done by other individuals. Because Dow Jones's Factiva was created for business clients, where there might be only one or two assigned users within a corporation, it allows users to set search preferences that remain in place until those preferences are changed (and saved). However, in a university environment, which typically only purchase one license with access to multiple 'seats' (to anyone with valid access to use the library system), this becomes challenging. As opposed to establishing preferences on the main search window (e.g. to include only news stories), which is standard across databases, if one person saves their user-generated preferences in a particular way (using a gear icon which opens up a hidden menu option), it affects the searches of everyone thereafter. This gear icon grants 'administrator access', much like a computer administrator has privileges to make certain changes to a computer in a computer lab that someone with only a standard login access might not have. This is particularly problematic in a university environment because library staff cannot prevent individual users from making such administrative level changes in Factiva, and as Driedger and Weimer (2015) discovered, many librarians were not even aware this access even existed.

In collecting traditional media data, Driedger (2007) advocates using the broadest keywords possible because this best allows researchers to use their own judgement about how they want to filter the data for what they consider 'useful' to their study. For example, in Driedger's study examining risk perceptions of drinking water after an E. coli contamination event where 7 people were killed and more than 2,300 people became seriously ill in Walkerton, Ontario, Canada, keywords like 'Walkerton' and 'drinking water' paradoxically eliminated stories that were relevant to the study purpose. The study was primarily interested in how people made sense of the Walkerton drinking water event. It was also interested in how people used the Walkerton drinking water contamination as an example of a type of risk a community (located anywhere) wanted to avoid, even if that risk was not specific to drinking water. It was equally concerned with how the different levels of government responded to the Walkerton event (new regulations, rules, or what might need doing to protect drinking water generally) as well as how critiques of a government's past performance (cutbacks, devolution of responsibilities, etc.) was portrayed in the media. Using the broad keyword 'Walkerton' allowed the researcher to be the filter of what stories were relevant to the study purpose; even if this meant having to go through many more stories. A narrower search like 'Walkerton' and 'drinking water' only yielded about one-third of possible stories to evaluate but excluded what the researcher felt were very relevant stories. Hence, as in any type of data collection issues, the tradeoffs that a researcher makes (time vs. comprehensiveness) is important and should be clearly justified within the study as well as in the limitations of the study. Last, incorporating strong inclusion and exclusion criteria are important to justify what and how aspects of the study are being operationalized so that others might be able to replicate the process. These can include date ranges, newspaper specific searches and types of story content that 'meet' study criteria. Good practice in traditional media studies also suggests that approximately 60% of the story should be about the topic of interest (Sampert and Trimble, 2009), however this approximation remains defined by the researcher.

By contrast, doing searches of new media can be more challenging (and more expensive). New media data involves both 'demand side' material on the Internet generated through search engine queries (websites, blogs) and 'supply side' material (the actual content published in those websites, blogs) (Eysenbach, 2009), where there are considerably fewer controls established about what kind of material is 'supplied'. Both Neeley (2014) and Batrinca and Treleaven (2015) offer broad groupings of social media into 'types' such as blogging, feeds, social networking media and even geospatial feeds (location metadata attached to social media data generated from mobile devices). Ultimately, any effort to categorize new media into distinct types of categories can create limitations for inclusion and exclusion criteria that any researcher needs to be attentive to when developing strategies for how broad or narrow they will cast their data collection net.

There are a number of challenges in accessing data from new media that is consistent and systematic. Unlike traditional media e-resource databases that offer the capacity to do keyword specific searches within a date range permitted by the available collection, new media searches require multiple creative strategies that can yield different results depending on the location of the computers being used or the search engine being accessed (Pariser, 2011; Carr, 2008). This trend, referred to as 'personalization', uses algorithms to track various signals – the computer being used, its location, the browser, and so forth – to shape the results that are returned based on user-initiated keywords (Pariser, 2011). As Neeley (2014, p. 154) writes:

> The problem with personalization deepens when the world of search results is narrowed even further, based on your browsing history and what you've 'liked' on Facebook or Google+. With frictionless sharing, this information not only influences your own experience, but it starts to shape what is presented and recommended to your friends as well. Since the rise of social media, one key concern has been whether online communities are insular 'echo chambers' that not only aggregate similar-minded people, but also then proceed to harden those views and even make them more extreme.

As Pariser (2011) urges during a TED talk, 'algorithms do not have embedded ethics'. In developing his argument, he identifies that in the turn of the last century, when newspapers were beginning, many journalists and editors did not have a set of professional ethics for how they practiced their craft. Yet, within a relatively short period of time, many realized that the news media provided a very important function in a democratic society, provided that there was a good flow of information within civil society. Because journalists and editors at newspapers were (and still are) acting as a filter, it was imperative that a set of professional conduct and ethics be established. However, Pariser (2011) argues that for the Internet and new media it is like we are back to the early days of newspapers:

> What we're seeing is more of a passing of the torch from human gatekeepers to algorithmic ones. So if algorithms are going to curate the world for us, if they're going to decide what we get to see and what we don't get to see, then we need to make sure that they're not just keyed to relevance. We need to make sure that they also show us things that are uncomfortable or challenging or important.

While search personalization raises concerns with what one actually is able to view, the new media social researcher runs the risk of being the equivalent of a 'drunk looking under the lamp post to find his keys because that's where the light is the brightest' (Slee, 2011). Embedded within these challenges is a host of issues that researchers need to grapple with when they conduct searches to access new media data for analysis. Accessing new media data can be achieved in multiple ways (e.g., Facebook, blogs, etc.), although there are considerable limitations with each.

To fully understand the inherent complexity and limitations of available systems for searching and retrieving new media data, a metaphor is useful. Consider the United States House of Representatives page system (appointed high school juniors that served the House), which was in operation on a national scale until 2011 (*Wikipedia*, 'United States House'). Pages were runners who were dispatched on a series of support tasks for the various House offices, but their most important job was retrieving and transporting documents of various types from multiple locations. Therefore, essential to the page function was access to information on the various floors of the Library of Congress, especially the restricted ones. What is relevant in this metaphor is that the page did not originate requests for information (i.e., the search terms), nor were they responsible to evaluate what was retrieved (i.e., the search finds) – this was the responsibility of whomever tasked the page with the list of documents to retrieve. Rather, pages were the delivery mechanism of that 'data' (i.e., the mechanical search engine) – the quality of which varied based on individual capacities such as speed, endurance and knowledge of and degree of access to the library to expedite information delivery.

The quality of House of Representative pages functions very much like the purpose of application programming interfaces (APIs) in the digital media world. When users go into a particular new media platform (Google, Facebook or Twitter), they rely on APIs to 'run' searches based on keywords. The outputs of these searches represent 'scrapes'. Scraping is both a technique and an analytic practice (Marres and Weltevrede, 2013). Scraping, defined, is 'the automated collection of online data . . . bits of software code that makes it possible to automatically download from the Web, and to capture some of the large quantities of data about social life that are available on online platforms like Google, Twitter and Wikipedia' (Marres and Weltevrede, 2013, p. 313). As a process, scraping not only locates data that meet user-specified parameters, but it also extracts data from a particular Internet page that meets precise criteria as opposed to being an indiscriminate extraction from the Internet (Marres and Weltevrede, 2013). That precision tends to be based on controls established in the background by the original programmers or established by those specific Internet pages about what outside users have access to or not.

However, the purpose of APIs varies greatly from platform to platform, and these are further varied based on aspects of personalization described above. The API can filter content to provide the user something that more closely matches that user's general preferences based on the user's digital footprint. Moreover, online platforms that produce APIs (automated background programming code that establishes boundaries for what kind of data can be accessed) also regulate the access to 'otherwise 'generally accessible' platforms . . . [which] may end up placing severe constraints on social research' (Marres and Weltevrede, 2013, p. 322). In the context of an API within a proprietary platform, like Twitter, Boyd and Crawford (2012, p. 669–70) write:

> It is not clear what tweets are included in these different data streams or [what] sampling them represents. It could be that the API pulls a random sample of tweets or that it pulls the first few thousand tweets per hour or that

> it only pulls tweets from a particular segment of the network graph. Without knowing, it is difficult for researchers to make claims about the quality of the data that they are analyzing. . . . Twitter has become a popular source for mining Big Data, but working with Twitter data has serious methodological challenges that are rarely addressed by those who embrace it. When researchers approach a data set, they need to understand – and publicly account for – not only the limits of the data set, but also the limits of which questions they can ask of a data set and what interpretations are appropriate.

Recognizing these constraints and limitations in a study design is important. However, while many present the idea of collecting new media data as 'simple and easy' (Yoon, et al., 2013, p. 127), this is only really true of those who are already comfortable with computer programming. Because new media is an ever changing environment, where published methodologies for mining data quickly become obsolete even though the steps to follow remain roughly the same (see Yoon, et al., 2013), digital media social researchers need to be creative. Batrinca and Treleaven (2015) provide a pretty thorough overview of options and strategies for collecting new media data. Time will tell for how long those strategies remain relevant.

At its most basic level, there are some strategies available to most non-computer-science-based researchers that range from the least to the most expensive. Going to specific social media sites and doing specific searches is one option. For example, Facebook is organized by *pages*, which are predominantly one-way sources of information with limited interactivity, and *groups*, which are member-based and more interactive and personal. The pages or groups selected for analysis on a given topic can be identified against a series of relevant parameters: the number of followers or 'likes' (for pages) or the number of members (for groups). Another strategy is to pay for a third party company (e.g., Gnip) that archive and resell new media data, like Twitter. However, these third party companies operate on a business model intended for larger organizations with a capacity to pay as opposed to individual academics with limited research grants or students having to pay the direct costs of research themselves. Hence, comprehensive historical searches can be very expensive with a wide pricing range given the number of keywords, geographical locations and so forth required for the search. For example, for one of Driedger's projects on chronic cerebrospinal venous insufficiency (CCSVI), popularized as 'liberation therapy', for people with multiple sclerosis, she received a quote of $20,000 to conduct a historical Twitter search for three to five keywords back in April 2013. Given the size of her grant and the commitment to other data collection needs, Driedger pursued using only Facebook (which was 'free' and could do historical page searches) to coincide with a traditional newspaper media analysis. Consequently, funding feasibility is a key criteria in academic environments for all aspects of any research project, when often, to be able to make use of any kind of data requires financial support to acquire the data or hire research assistants to help with the analysis. Nonetheless, platforms like Twitter, for example, are making efforts to change this. In February 2014, Twitter announced its Data Grants Program to 'give a handful of research institutions access to our public and historical data' (Twitter, 2014); however, it is presently not accepting submissions (Twitter, n.d.).

The strength of new media is often its inherent limitation: while content can be added to, changed, challenged, or promoted, which enables a more comprehensive understanding of how users of those media shape and define issues of concern, aside from being very expensive, it typically does not facilitate historical examination. Hence, unless you are actively following issues in real time and collecting that data, it is not always easily retrieved at a later date. Notably, Batrinca and Treleaven (2015) also identify this as a problem for real-time data collection. By contrast, despite its limitations for analyzing social life, traditional media data sets are inexpensive to use, have an established set of ethical conduct and responsibility that guides how content is generated, and enable both retrospective and prospective data collection. That said, some question the relevance of traditional media to new generations of news consumers (Diddi and LaRose, 2006; Tewksbury, 2003); although it is possible that new generations of news consumers still 'read' traditional media depending on how traditional story items may be trending on new media platforms.

Analyzing and reporting on media findings

Regardless of the dominant methodology underpinning a study – quantitative, qualitative, or mixed – there are a number of common features that are included in the reporting of the data. 'Quantitative' analyses include reporting of the number of stories and items or trends and patterns over time for more longitudinal examinations. From a 'qualitative' perspective, there is often an examination of what is said, how is it said, and who or what is noticeably absent.

Many quantitative analyses of media data typically rely on content analysis which tends to emphasize frequencies (Driedger, 2007; Kennedy and Bero, 1999; Wenger, et al., 2001; Durrant, et al., 2003; Harris, et al., 2012; Leccese, 2009; Buchanan, 2014; Keelan, et al., 2010; Love, et al., 2013) but sometimes involves more sophisticated statistical approaches (Armstrong and Boyle, 2011; Haigh and Heresco, 2010). Anytime a researcher wants to document how much attention is given to an issue, how media attention might increase or decrease the emphasis of an issue over time, a quantitative content analysis is frequently used (see Krippendorff, 2004). While Patton (2002) has also made reference to some approaches to qualitative analysis as a form of content analysis, important distinctions are needed for media analyses that help to separate quantitative treatments of data from qualitative ones. A quantitative content analysis typically focuses on manifest content – the surface descriptive content of the words used in the presentation of a media story. In such circumstances, there is more emphasis on the number of stories, the length of stories, the placement of stories and so forth, and where coding guides for capturing lines of text are often generated a priori.

By contrast, a qualitative analysis of media presentations of an issue might adopt several different strategies. There might be an initial coding of data according to manifest content categories, and in the presentation of results, may include numeric counts of terms, concepts or themes. While counting occurrences is useful for establishing the magnitude of a concept or term, it lacks deeper understanding. Therefore, much of the emphasis in a qualitative analysis extends this to focus its examination more on the latent content, or the interpretive meaning behind the written text (e.g., see Deacon, et al., 1999; Driedger and Eyles, 2003; Klowdawsky, Farrell and D'Aubry, 2002; Driedger, et al., 2014; Porto and Romano, 2013; Myers, et al., 1996; Harding, 2006; Clarke and Mosleh, 2014; Lehman, et al., 2013). There are different strategies that can be used in such qualitative analyses (e.g. Braun and Clarke, 2006) and common among these is content analysis informed by framing theory. A qualitative content analysis informed by framing theory does not treat each utterance of words the same. If there is a content category present in the lines of text that provides little substance beyond passing statements, they are typically not included in the results. Rather, only those aspects that contribute to the salient frames of the text – that is, contributing to a greater interpretive read of the text – is considered as part of the data for treatment in a qualitative study. Entman writes (1993, p. 57), 'Unguided by a framing paradigm, content analysis may often yield data that misrepresent the media messages that most audience members are actually picking up'. Thus, in a framing informed analysis, the emphasis is on the meaning that most readers might actually notice in the text.

For example, within such framing analyses, there might be an examination of how sources try to shape the definition of a problem or issue, albeit still filtered by how the media presents those statements (e.g. Driedger, 2008; Klodawsky, et al., 2002). Likewise, there might be comparative analyses being conducted to examine how well dominant frames contained in the story lead matches dominant frames in the body of the story (e.g. Mistry and Driedger, 2012), how the initial framing of an issue in the early days may remain unchanged over time (Driedger, et al., 2009) or how online comments can be examined to challenge dominant story frames within traditional news media stories (e.g., Muzyka, et al., 2012). Framing, thus, is a way to capture and understand dominant discourses, counter-discourses and transgressive views in a theoretical space similar to Foucault's 'heterotopia' (Foucault, 1986).

Ultimately, the way in which media data sets are analyzed depends on the research questions guiding the study. Likewise, the way in which findings are reported also needs to have a good fit for purpose. Nonetheless, in any analysis, operationalizing how units of text are being coded for analysis into relevant content categories or categories of meaning is important, and seeking guidance from how others in the literature have described their own strategies are useful starting points to help guide any future study.

Any section on analyzing data would be incomplete without recognizing the burgeoning use of computer assisted qualitative data analysis software (CAQDAS) over the last twenty plus years. While many different software packages are available, the authors of this chapter are most familiar with the use of QSR International's NVivo and only provide some brief comments for that software. Excellent resource tools for working with qualitative data (Richards, 2014) and using NVivo (Bazeley and Jackson, 2013) are highly recommended. And although NVivo10 has been made available for Mac users, as opposed to its prior exclusive support for Windows users, researchers should familiarize themselves with continued limitations for Mac users before making a purchase decision (QSR International, 2015). Anecdotally, Driedger's students who have used Mac versions of NVivo10 have found serious limitations in its functionality with respect to social media data, suggesting that the software still has a way to go to be equally comparable to NVivo's Windows-based version.

With traditional news media data, downloaded stories (however managed in the database) can be brought into NVivo much like you would an interview transcript Word document. Based on Driedger's experience, importing each individual story as its own Word document, while a bit more time consuming, is the preferred option over bringing in stories from particular newspapers for specific months or years. In this way, each story can be coded as its own case node upon import and assigned relevant attribute information by the researcher. Likewise, if you have taken the time to structure that document using the header functionality in Word (e.g., Header 2), you can organize aspects of the story into building blocks of meaning relevant to the research questions through autocoding. For example, in Mistry and Driedger (2012), individual stories were structured in Word to separate the lead, body, and conclusion using Header 2, thereby autocoding these and other categories relevant to those authors for further analysis through discretionary researcher-determined coding. While manual coding these broad types of sections can be done, when working with a large data set, inserting relevant organizational category labels in a Header 2 style was relatively easy to do while performing other data clean-up typically required when downloading stories from an e-resource database like Proquest's Canadian Newsstand Major Dailies.

With specific relevance to new media, the most recent version, NVivo 10, has included functionality through NCapture[2] that enables direct capture of data from web pages, social media conversations and Evernote. (Note: all content that follows is based on performance of NVivo10 for Windows). In the case of Web Pages, NCapture automatically converts this content to a PDF file and facilitates the coding of the entire page to one or more nodes at the moment of 'capture'. After the web page PDF is imported into your NVivo project (through External Data, NCapture), more discretionary coding can take place, much like you would with a PDF document. With respect to social media platforms, NVivo 10 is designed to work easily with Facebook, Twitter and LinkedIn, again using the NCapture functionality, but where the content is brought in as a data set. When using NCapture for Facebook, the posts, number of likes, public information about the profile of the poster and any comments are displayed. The autocode features of NVivo 10 can allow for the conversations to be autocoded and for attributes to be created

2 NCapture is an add-on to be installed for Internet Explorer (version 8 or later) and Chrome (version 21 or later).

for each individual contributor based on publicly available information and so forth to facilitate future queries of the data. For Twitter, searching can be done by phrases or hashtags, or by searching for tweets by individuals or organizations, and NCapture can bring in the desired selections. Again, autocoding functionality can create individual cases for each tweeter, hashtag or location, as well as charts for how tweet activity has changed over time, in addition to more discretionary coding done by the individual researcher. With Evernote, you simply export the specific notes you wish to bring into NVivo, import the external saved file (external data, from Evernote) and code, develop memos and run queries as relevant to your needs. However, as Bazeley and Jackson (2013, p. 245) wisely point out, the results from any user-generated NVivo queries 'will only be as good as is allowed by the combination of your skill in coding, linking and reflecting within the database, your ability to ask relevant questions (to run appropriate queries), and your capacity to interpret – and challenge – the output.'

Ethical challenges

Traditional media such as print, television and radio present few ethical challenges in terms of access to data, conducting an analysis and reporting findings. Almost all traditional media sources publicly release data in the form of publications or newscasts. As publicly available information, ethics approval is not required to analyze these sources. By comparison, new media sources can be an ethical minefield. Most important to consider when collecting new media is the public or private nature of the information and an individual's assumption of privacy. Consider, for example, individuals posting on Facebook but making comments regarding a publicly available news story. If someone makes his or her profile private (in that others are required to 'friend' them in order to see the information), there is an ethical assumption of privacy despite what we may all know about lack of real, or actual, privacy. In this situation, ethics approval would be required to include such sites in a research project, and all ethical requirements would apply. These would likely include full disclosure of the purpose and intent of the work and informed consent from the page owner. However if a page is public, most ethics boards would suggest that there is the assumption of public observation, and ethics approval is not required. Therefore, when the focus of a new media study is to simply analyze the content posted by others in a public forum or page, research ethics approval is not normally sought, just as ethics approval would not be required to observe behavior in a public place such as a park or sidewalk.

When one is required to join a group or be a 'friend' (for example), ethical considerations get more complex and REB approval should be sought. The degree of ethics oversight, and required safeguards, will depend on the research. It is important to understand the goal of your ethics board is to protect the rights and dignity of research participants. Therefore we always suggest that if there are any doubts as to whether ethics approval is required, you should consult with your own authority. A standard rule is: 'When in doubt, get approval.'

For new media research that is low-risk (where participants would not experience risk greater than that in everyday life), most ethics boards permit the creation of a real or fictional person account (something that is often needed to view or like the content of certain group sites on social media) provided that the researcher does not contribute to the discussion dialogue in any way. An example might be the creation of a Twitter account to follow discussions on a particular topic of interest. Hence, examination of new media discourse as a 'voyeur' is generally permitted, so long as the researcher is not an active participant. Ethics boards will also ask that researchers preserve the anonymity of individuals posting to these sites.

Where new media research is of higher risk, or where there might be some personal risk of exposure or personal cost to being identified, ethics oversight is concomitantly higher. For example, should a researcher wish to investigate a Reddit thread related to teen girls and body image, ethics approval should be sought because the subject of the research includes minors, which raises the risk classification of the research. If simple observation is sought, most ethics boards are likely to approve. However if the researcher plans to create a profile and 'pretend' to be one of the teen girls, this classifies the research as 'including deception' and again moves the project into a higher risk category, requiring additional steps such as approval to join the group and debriefing. As stated previously, researchers should consult with their local ethics boards prior to conducting research.

When planning research projects, including investigations of new media sites, the protection of participants is paramount. Even though sites may be publicly available it does not grant researchers the right to infringe on the privacy and confidentiality of an individual by revealing identities. In this way, participants via new media sources are no different than any other participant. Even with traditional media analyses, researchers typically protect the identities of locals within their manuscripts by referring to the specific grassroots organization or by blocking certain identifying information within a quotation. While these measures are not perfect – anyone really interested can go look up a specific site or traditional story – it takes greater effort for an outside person to obtain the identity being portrayed within academic research. Lastly, university REB offices may not have formal policies with respect to new media, and hence, it can also be challenging to obtain permission to study certain content for research purposes. In these cases it is best to consult your ethics board early in the research design process to prevent last-minute surprises and time delays.

Conclusion

Media analysis is a powerful tool for identifying the framing of ideas, the tracking of issue importance, and identification of who has (or has not) the power and credibility to speak on a particular topic. Traditional media can shape public discourse by defining what is important, and by framing the context in which public discussion can take place. New media serve an equally important function in providing an arena for alternative and conflicting viewpoints. Where communication in traditional media is highly controlled, it is rarely controlled in new media. In this way, traditional media often provide examples of top-down framing, whereas new media provide examples of bottom-up framing. Wherever possible, health geographers would be best served by including both in research projects in the future.

References

Abramson, K., Keefe, B., and Chou, W. S., 2015. Communicating about cancer through Facebook: a qualitative analysis of a breast cancer awareness page. *Journal of Health Communication: International Perspectives*, 20(2). pp 237–243.

Anonymous, undated. Data Grants Closed. Engineering Blog. https://engineering.twitter.com/research/data-grants-closed/. Accessed March 4, 2015.

Armstrong, C. L., and Boyle, M. P., 2011. Views from the margins: news coverage of women in abortion protests, 1960–2006. *Mass Communication and Society*, 14(2). pp. 153–177.

Batrinca, B., and Treleaven, P. C., 2015. Social media analytics: a survey of techniques, tools and platforms. *AI and Society*, 30(1). pp. 89–116.

Bazeley, P., and Jackson, K., 2013. *Qualitative Data Analysis with NVivo*. 2nd ed. Los Angeles: Sage Publications.

Braun, V., and Clarke, V., 2006. Using thematic analysis in psychology. *Qualitative Research in Psychology*, 3 (2). pp. 77–101.

Boyd, D., and Crawford, K., 2012. Critical questions for big data: provocations for a cultural, technological and scholarly phenomenon. *Information, Communication & Society*, 15(5). pp. 662–679.

Boyd, A. D., Jardine, C. G., and Driedger, S. M., 2009. Canadian media representations of mad cow disease. *Journal of Toxicology and Environmental Health*, 72(17–18). pp. 1096–1105. doi: 10.1080/15287390903084629.

Buchanan, C., 2014. A more national representation of place in Canadian daily newspapers. *The Canadian Geographer*, 58(4). pp. 517–530.

CIHR (Canadian Institutes of Health Research), NSERC (Natural Sciences and Engineering Research Council of Canada), and SSHRC (Social Sciences and Humanities Research Council of Canada), 2014. *Tri-Council policy statement: ethical conduct for research involving humans*. Available at: http://www.pre.ethics.gc.ca/pdf/eng/tcps2-2014/TCPS_2_FINAL_Web.pdf.

Carr, N., 2008. Is Google making us stupid? *The Atlantic*. Available at http://www.theatlantic.com/magazine/archive/2008/07/is-google-making-us-stupid/6868/.

Clarke, J. N., and Mosleh, D., 2014. Risk and the Black American child: representations of children's mental health issues in three popular African American magazines. *Health, Risk & Society*, 17(1). pp. 1–14.

Cohen, B., 1963. *The Press and Foreign Policy*. Princeton, NJ: Princeton University Press.

Deacon, D., Fenton, N., and Bryman, A., 1999. From inception to reception: the natural history of a news item. *Media Culture and Society*, 21(1). pp. 5–31.

Diddi, A., and LaRose, R., 2006. Getting hooked on news: uses and gratifications and the formation of news habits among college students in an Internet environment. *Journal of Broadcasting & Electronic Media*, 50(2). pp. 193–210.

Driedger, S.,M., and Weimer, J., 2015. Factiva and Canadian Newsstand Major Dailies: comparing retrieval reliability between academic institutions. *Online Information Review*, 39(3). pp. 346–356. doi:10.1108/OIR-11-2014-0276.

Driedger, S. M., 2007. Risk and the media: a comparison of print and televised news stories of a Canadian drinking water risk event. *Risk Analysis*, 27(3). pp. 775–786.

Driedger, S. M., Jardine, C., Boyd, A., and Mistry, B., 2009. Do the first ten days equal a year? comparing two Canadian public health risk events using the national media. *Health, Risk and Society*, 11(1). pp. 39–53. doi: 10.1080/13698570802537011.

Driedger, S. M., Mazur, C. and Mistry, B., 2014.The evolution of blame and trust: an examination of a Canadian drinking water event. *Journal of Risk Research*. 17(7–8). pp. 837–854. doi:10.1080/13669877.2013.816335.

Driedger, S. M., 2008. Creating shared realities through communication: exploring the agenda-building role of the media and its sources in the E. coli contamination of a Canadian public drinking water supply. *Journal of Risk Research*, 11(1–2). pp. 23–40.

Driedger, S. M., and Eyles, J., 2003. Different frames, different fears: Communicating about chlorinated drinking water and cancer in the Canadian media. *Social Science and Medicine*, 56. pp. 1279–1293.

Durfee, J., 2006. 'Social change' and 'status quo' framing effects on risk perception: an exploratory experiment. *Science Communication*, 27(4), pp. 459–495.

Durrant, R., Wakefield, M., McLeod, K., Clegg-Smith, K., and Chapman, S., 2003. Tobacco in the news: an analysis of newspaper coverage of tobacco issues in Australia, 2001. *Tobacco Control*, 12(2). pp. 75–81.

Entman, R., 1993. Framing: toward clarification of a fractured paradigm. *Journal of Communication*, 43. pp. 51–8.

Eysenbach, G., 2009. Infodemiology and infoveillance: framework for emerging set of public health informatics methods to analyze, search, communication and publication behavior on the Internet. *Journal of Medical Internet Research*, 11(1), p. e11. doi:10.2196/jmir.1157.

Fleras, A., 2003. *Mass media communication in Canada*. Scarborough, ON: Thomson-Nelson.

Foucault, M., 1986. Of Other Spaces. *Diacritics*, 16 (1). pp. 22–27.

Gamson, W. A., 1989. News as framing comments on Graber. *The American Behavioral Scientist (1986–1994)*, 33(2), p. 157

Harding, R., 2006. Historical representations of aboriginal people in the Canadian news media. *Discourse & Society*, 17(2). pp. 205–235.

Harris, A. J., Gurioli, L., Hughes, E. E., and Lagreulet, S., 2012. Impact of the Eyjafjallajökull ash cloud: a newspaper perspective. *Journal of Geophysical Research*, 117 (B9), doi:10.1029/2011JB008735.

Haigh, M. M., and Heresco, A., 2010. Late-night Iraq: monologue joke content and tone from 2003 to 2007. *Mass Communication and Society*. 13(2). pp. 157–173.

Iyengar, S. 1991. *Is anyone responsible? How television frames political issues*. Chicago: University of Chicago Press.

Keelan, J., Pavri, V., Balakrishnan, R., and Wilson, K., 2010. An analysis of the Human Papilloma Virus debate on MySpace blogs. *Vaccine*. 28(6). pp. 1535–1540.

Kennedy, G., and Bero, L., 1999. Print media coverage of research on passive smoking. *Tobacco Control*, 8. pp. 254–260.

Kim, S. H., and Anne Willis, L., 2007. Talking about obesity: News framing of who is responsible for causing and fixing the problem. *Journal of health communication*, 12(4), pp. 359–376.

Klodawsky, F., Farrell, S., and D'Aubry, T., 2002. Images of homelessness in Ottawa: implications for local politics. *The Canadian Geographer*, 46(2). pp. 126–143.

Krippendorff, K., 2004. *Content analysis: an introduction to its methodology*, 2nd ed. Thousand Oaks, London, New Delhi: Sage Publications.

Lang, G. E., and Lang, K., 1981. Watergate: an exploration of the agenda-building process. *Mass Communication Review Yearbook*, 2. pp. 447–468.

Leccese, M., 2009. Online information sources of political blogs. *Journalism & Mass Communication Quarterly*, 86(3). pp. 578–593.

Lehmann, B. A., Ruiter, R., and Kok, G., 2013. A qualitative study of the coverage of influenza vaccination on Dutch news sites and social media websites. *BMC Public Health*, 13(1), p. 1. doi:10.1186/1471-2458-13-547.

Lippmann, W., 1922. *Public opinion*. New York: Free Press.

Love, B., Himelboim, I., Holton, A., and Stewart, K., 2013. Twitter as a source of vaccination information: content drivers and what they are saying. *American Journal of Infection Control*, 41(6). pp. 586–570.

Marres, N., and Weltevrede, E., 2013. Scraping the social? Issues in live social research. *Journal of Cultural Economy*, 6(3). pp. 313–335.

McCombs, M. E., and Shaw, D. L., 1972. The agenda-setting function of mass media. *Public Opinion Quarterly*, 36. pp. 176–87.

McCombs, M. E., and Shaw, D. L., 1993. The evolution of agenda-setting research: twenty-five years in the marketplace of ideas. *Journal of Communication* 43. pp. 58–67.

Mistry, B., and Driedger, S. M., 2012. Do the leads tell the whole story? An analysis of story leads of the Walkerton, Ontario, E. coli contamination of drinking water supplies. *Health Risk and Society*, 14(6). pp. 583–603. doi: 10.1080/13698575.2012.701275.

Muzyka, C. M., Thompson, L. H., Bombak, A. E., Driedger, S. M. and Lorway, R., 2012. A Kenyan newspaper analysis of the limitations of voluntary medical male circumcision and the importance of sustained condom use. *BMC Public Health*, 12(1), p. 1. doi: 10.1186/1471-2458-12-465.

Myers, G., Klak, T., and Koehl, T., 1996. The inscription of difference: news coverage of the conflicts in Rwanda and Bosnia. *Political Geography*, 15(1). pp. 21–46.

Neeley, L., 2014. Risk communication in social media. In Arvai, J., and Rivers, L., III, (eds.) 2014. *Effective Risk Communication*. pp. 143–164. New York: Earthscan from Routledge.

Pariser, E., 2011. Beware online 'Filter Bubbles'. (Ted Talks) *TED2011*. Available at http://www.ted.com/talks/eli_pariser_beware_online_filter_bubbles.html/.

Patterson, T. E., and Donsbac, W., 1996. News decisions: journalists as partisan actors. *Political Communication*, 13. pp. 455–68.

Patton, M. Q. 2002. *Qualitative research and evaluation methods*. 3rd ed. Thousand Oaks, CA: Sage.

Porto, D., and Romano, M., 2013. Newspaper metaphors: reusing metaphors across media genres. *Metaphor and Symbol*, 28(1). pp. 60–73.

Price, V., and Tewksbury, D., 1997. News values and public opinion: a theoretical account of media priming and framing. *Progress in communication sciences*, pp. 173–212.

QSR International, 2015. NVivo feature comparison – NVivo 10 for Windows and NVivo 10 for Mac. Available at http://www.qsrinternational.com/product-comparison/nvivo10-nvivo-for-mac.html. Accessed August 20, 2015.

Ramo, D. E., Liu, H. and Prochaska, J. J., 2015. A mixed-methods study of young adults' receptivity to using Facebook for smoking cessation: if you build it, will they come? *American Journal of Health Promotion*, 29(4). pp. e126–e135.

Richards, L., 2014. *Handling qualitative data: A practical guide*. 3rd ed. London: Sage.

Rogers, E. M., and Dearing, J. W., 1988. Agenda-setting research: where has it been? Where is it going? In Anderson, J. A. (ed.), 1988. *Communication Yearbook 11*. Newbury Park, CA: Sage. pp. 555–594.

Sampert, S., and Trimble, L., 2009. *Mediating Canadian politics*, Toronto: Pearson Press.

Scheufele, D. A., 1999. Framing as a theory of media effects. *Journal of Communication*, 49(1). pp. 103–122.

Scheufele, D. A., and Tewksbury, D., 2007. Framing, agenda setting, and priming: the evolution of three media effects models. *Journal of Communication*, 57(1). pp. 9–20.

Slee, T., 2011. Internet-centrism 3 (of 3): tweeting the revolution (and conflict of interest). September 22, [blog]. Available at http://whimsley.typepad.com/whimsley/2011/09/earlier-today-i-thought-i-was-doomed-to-fail-that-part-3-of-this-prematurely-announced-trilogy-was-just-not-going-to-get-wr.html accessed March 11, 2015.

Soroka, S., 2002. *Agenda-setting dynamics in Canada*. Toronto: UBC Press.

Tankard, J., Hendrickson, L., Silberman, J., Bliss, K., and Ghanem, S., 1991. Media frames: Approaches to conceptualization and measurement. Paper presented to the Association for Education in Journalism and Mass Communication, Boston.

Tewksbury, D., 2003. What do Americans really want to know? Tracking the behavior of news readers on the Internet. *Journal of Communication*, 53(4). pp. 694–710.

Twitter, 2014. Introducing Twitter data grants. Available at: https://blog.twitter.com/2014/introducing-twitter-data-grants. Accessed February 5, 2016.

Twitter, n.d. Data grants closed. Available at: https://engineering.twitter.com/research/data-grants-closed. Accessed February 5, 2016.

Yoon, S., Elhadad, N., and Bakken, S., 2013. A practical approach for content mining of tweets. *American Journal of Preventive Medicine*, 45(1). pp. 122–129.

Wenger, L., Malone, R., and Bero, L., 2001. The cigar revival and the popular press: a content analysis, 1987–1997. *American Journal of Public Health*, 91(2). pp. 288–291.

Wikipedia. United States House of Representatives page. Available at http://en.wikipedia.org/wiki/United_States_House_of_Representatives_Page. Accessed March 22, 2015.

Part IV

(Non)representation, affect and social life

12 From 'The pump' to 'Senescence'

Two musical acts of more-than-representational 'acting into' and 'building new' life

Gavin J. Andrews and Eric Drass

Reflections on praxis

This chapter describes collaboration between a health geographer and musician, which involved making pop music in academic contexts. The challenges that emerged fall broadly under two themes. First, there was the challenge of doing non-representational theory itself. Of overcoming the academic urge to 'dig down' and theorize phenomena, and instead engage with public life more lightly and directly. Also of not employing 'safe' tried and tested qualitative methodological data collection approaches, but instead producing something new: sounds that proactively 'act into', 'change' and 'boost' the world a little. Second, there was the challenge of working with very different types of professionals and their mediums (a researcher with an artist and their sounds, and an artist with an academic and their academic words). Indeed, fitting music into scholarship and scholarship into music was not problem-free, and similarly understanding one another and one another's motivations and approaches was an issue. Such collaboration can be tricky due to conflict with personal and professional conventions, but the result was, for us, rewarding, bold and impactful. The chapter provides a snapshot of future research if non-representational theory were to establish itself further in health geography in the way that it has in the parent discipline; a future where praxis is fundamental because it involves the creation of new realities.

Introduction

Non-representational theory (NRT) is gradually gaining ground and being deployed in health geography. One of the main facets of the approach yet to be considered in any great depth in the subdiscipline, is the methodological creation of new realities that 'act into' the world to boost, change or build new forms of life. Providing some ideas on how this radical way of dealing with subjects and of doing research might arise, this chapter reports on two engagements between a multimedia artist and health geographer, the former as the musical writer, composer and producer, the latter as a supporter, occasional sounding board and subject. The first, titled 'The Pump' – a high energy industrial techno track about extreme fitness and body cultures – was more accidental, emerging in 1998 whilst both authors were living together with friends and starting their respective careers. The second, titled 'Senescence' – an ambient electronic track about biological aging – was more purposefully conceived as an academic project 14 years later in 2012 with specific artistic and geographical purposes. The analysis covers the nature of the engagements and of the two products, ranging from personal, creative, technical and structural features to the production of affect and meaning in the two soundscapes, and their circulation. The chapter rounds off with some final thoughts on the broader challenges and opportunities of producing and working with art in health geography and releasing vibes into the world within a more-than-representational paradigm.

Enter non-representational theory

Given all the attention it has received over the last few years across numerous published books, papers and chapters, one would have had to have been stranded on an island as a qualitative human geographer to have missed all the fuss about NRT (see, for example, Lorimer, 2005; 2007; 2008; Thrift, 2008; Cadman, 2009; Anderson and Harrison, 2010; Vannini, 2014a). Originally emanating from the UK, but now gaining broader international appeal, as an overall approach NRT introduces a new way of going about the business of contemporary human geography. It is based on the observation that a sizeable part of the world and life – what actually happens 'out there' in space-time – has either been completely ignored or deadened in the process of analysis and representation by mainstream social constructivist qualitative research (Andrews, et al., 2014a). This is attributed to social constructivism's deep commitment to theoretically driven interpretative searches for power, meanings, emotions and significance, and also to the strict orders, structures and processes that are imposed by researchers who employ it (Dewsbury, et al., 2002). In contrast, NRT does not see an external world waiting to be theorised away by detached observers. The idea is instead that they engage the lived world as an ongoing and performative achievement, and they present life in its lively, immediate and continual moving form (Thrift, 2004). This involves understanding action to be at the forefront – rather than as a background – of peoples' lives (Boyd and Duffy, 2012), and attention being paid to the numerous modest, unspoken, spontaneous and often involuntary practices that together make life. This is so that, as Thrift (2008) explains, research might reverberate how life shows up, expresses and feels in its most basic forms, often prior to it being fully cognitively realized (Thrift, 1997).

As Andrews (2014a) explains, beyond this, the main facets that together create and characterize NRT as a particular 'style' of research include a relational materialist positioning and specific interests in the rolling out and reveal of space-time, senses and sensations, practice and performance, vitality, virtuality and multiplicity, and the mundane and ordinary. Moreover, as we shall see later, NRT includes methods, theories and styles of intervening in and communicating with the world (each of which have been the subject of in-depth and focused attention in their own right – see Cadman, 2009; Dewsbury, et al., 2002; Thrift, 2000; Vannini, 2009;

2014a). These facets are certainly not all new or unique to non-representational theory and arise in other movements and traditions across the social sciences and humanities (including post-humanism, post-phenomenology, sensory ethnography, auto-ethnography, material culture studies, science and technology studies and performance studies). NRT however brings them together in a cohesive package and general approach; a potential new social science paradigm.

This said, it would be a stretch to claim that NRT constitutes a new paradigm specifically in qualitative health geography, a field that has traditionally followed, rather than led, theoretical developments in its parent discipline; a 'magpie' sub-discipline that borrows theory and ideas established elsewhere to help its own theory building (Kearns and Moon, 2002). However if one spreads the net wide enough – at one level classifying health geography as both research by health geographers and health research by other brands of human geographers, and at another level including health research that implicitly aligns well with some of the key facets of non-representational theory or mixes representational and non-representational approaches – the emergence of a 'more-than-representational' research orientation might be traced (see also Lorimer, 2005). In empirical research this has included a focus on relational feels and performances in range of community-based therapeutic and enabling situations (see Andrews, et al., 2014a; Barnfield, 2015; Tucker, 2010; Duff, 2011; 2012; Foley, 2011; 2014; Gatrell, 2013; Kraftl and Horton, 2007; Lea, et al., 2014; Macpherson, 2008; 2009a; 2009b; Philo, et al., 2014), much specifically in relation to the arts and care (Anderson, 2002; 2006; Andrews, 2014b; 2014c; Andrews et. al, 2013; Andrews and Shaw, 2010; Atkinson and Rubidge, 2013; Atkinson and Scott, 2015; Bissell, 2010; Evans et al., 2009; Evans, 2014; Greenhough, 2006; 2011a; 2011b; Greenhough and Roe, 2011; Jackson and Neely, 2014; McCormack, 2003; 2013; Simpson, 2014; Paterson, 2005; Solomon, 2011). Notably across much of this empirical research – as in much non-representational research in human geography more generally – the explanatory concept of 'affect' has been deployed frequently to illustrate the centrality and importance of sensations and basic feelings in health and care (Andrews, et al., 2013; 2014a; Andrews, 2014b; 2014c; Atkinson and Scott, 2015; Bissell, 2010; Boyer, 2012; Duff, 2010; 2011; 2012; 2014; Foley, 2011; 2014; Kraftl and Horton, 2007; McCormack, 2003; Philo, et al., 2014). Affect is a capacity and transitioning of a body, a process whereby it can be influenced by other bodies, modified in alignment with them, and can affect other bodies. While, the 'affective environment' means being affect's collective manifestation, something that is 'transhumanly' created (i.e., co-produced by and beyond humans) or 'transpersonally' experienced (i.e., co-shared between bodies prior to the personal) in space-time (Andrews, 2014a).

Theoretically, three notable developments have occurred in this literature over the past six or so years. First, actor-network theory – or at least a relational materialist position – has been used to expose the co-equal roles of humans and technologies in health and health care (Andrews, et al., 2013; Duff, 2011; Greenhough, 2006; 2011; Hall, 2004; Timmons, et al., 2010). Second, there has been some in-depth reflection on the philosophical foundation for more-than-representational health geographies that exists in the work of distinguished scholars, such as Georges Canguilhem's vitalism (Philo, 2007), Gilles Deleuze's transcendental empiricism (Duff, 2010; 2014), and Michel Foucault's College de France lecture series (Philo, 2012). Third, there has been some slightly more pragmatic discussion and debate surrounding the potential 'wins' and 'losses' of NRT as a new subdisciplinary direction in health geography, including potential foci, the need for continued caring and compassion for diverse client groups and 'relevance' to policy and practice (see Andrews, et al., 2014a; Kearns, 2014; Hanlon, 2014; Andrews, 2014a; 2015). Whilst these theoretical and positional contributions are welcome, health geography has failed to come to grips with the methodological changes, challenges and opportunities NRT poses for the subdiscipline.

More-than-representational methodological priorities

One fundamental methodological objective of NRT is, through methodological engagements, to 'witness' life. Specifically 'witnessing' is to pay close attention to the unfolding of space-time, including the numerous lively small happenings that together create it (see Dewsbury, 2003; Latham and Conradson, 2003). By doing this, it is hoped that the emerging data might have a fidelity and faithfulness to events. As Latham (2003) argues, existing methods do not have to be abandoned; they just need to be adjusted, augmented or combined 'to dance' a bit more with such techniques as diaries, go-along-interviews, experimental and expressive writing, photographs and video (Latham, 2003; Laurier, 2005; Lorimer, 2006; Wylie, 2005).

Another fundamental methodological objective of NRT is to 'act into' life (Andrews, 2014a; Thrift, 2008), which denotes a close relationship between the researcher and what is happening in the field and blurs the role of observer and what is observed (Dirksmeier and Helbrecht, 2008). Indeed with non-representational theory, method is itself a performance that does not so much as study a social reality through the acquisition of data as much as 'do' a social reality and live the data (Vannini, 2014a). For example, in addition to using some of the methods noted above, interviewing can be more about the interaction than the topics discussed, and participant observation can be reversed to 'observant participation,' which involves doing the same thing as the subject, getting more embroiled and invested in the effort and experience, and actively changing the course of events (Dewsbury, 2009; Thrift, 2000).

At the heart of NRT lies a political motivation to rally against and challenge the ways in which peoples' lives are manipulated by institutions under authoritarian neoliberal democracy and advanced capitalism. Thus a final methodological objective is to 'change' and 'boost' the active world, helping it to 'speak back'. These are all about introducing new realities into life, both in the field and in forms of knowledge translation. This is an incredibly important and somewhat unique methodological priority of NRT (Vannini, 2014a), not always part of every study, but seen as an ethical action in itself. The agenda towards 'building', 'changing', 'boosting' and 'speaking back' has been partly served by some of the methods mentioned above but also, quite specifically, by the development of arts-based research (often including theater and dance) and a more general agenda towards an activist and public scholarship.

More-than-representational methodological challenges

Importantly, underlying these methodological objectives, and absolutely critical to their success, is a disposition on the part of researchers, for them to approach the world with a sense of 'wonderment' (Thrift, 2008; Vannini, 2009; 2014a). It is critical for them to have a real sense of joy, appreciation and excitement to see, and participate in, all the movement and energy in life. This is because they must be able to look at places, events and situations through eyes unburdened by immediate thoughts of where social structures might be at work and where divisions might be occurring. They must be willing to take in the detail, feel the feels involved, unsettle and disrupt at the surface, and ultimately look to animate and reverberate these to their audience (Vannini, 2014a). Notably this sense of wonderment is not unlike that found in the popular spirituality movement generally and mindfulness techniques more specifically, whereby one might experience brief moments of 'self-transcendence' in observing, touching and listening to one's immediate environment. Experiencing it, and one's place in it, physically in the purest of forms, calmly as free from preconceptions and judgments as possible, and free from the constant stream – 'the madness' – of the mind's never-ending self-narrative. But it is also wonderment with the physicality of life that extends to greater scales. Thus not unlike the wonderment associated with theoretical physics – or at least its public articulation and popular imagining – it involves appreciating the almost endless physical threads and energies that span and bridge vast time and space to make life as it appears to us now, from atoms and molecules to complex bodies and objects. As Thrift (2008) notes, the opening for wonderment comes partly from the freedom gained by no longer having to fully explain or thoroughly theorize the social subject and from not having the troubles and injustices of the social world always at the fore in one's studies. Indeed, with these preoccupations and responsibilities removed from the forefront, the researcher has the space to be engrossed on a more basic level.

With NRT, two other methodological challenges have persisted that have been hotly debated, both concerned with the nature of happenings and feels, and the limitations of words and language in conveying them. The first arises because the focus of study is often a sensory experience produced from and inclusive of a wide range of body and object movements and interactions in places. Indeed, as Ducey (2007) suggests, full environmental and sensory happenings in life can never be translated directly through language and words – either on the part of subjects or researchers – which tend to simplify and deaden them or go too far down contemplative and interpretative paths in attempting to decipher them (Latham, 2003). The second challenge arises from the object of study – and later presentation – often being less than fully consciously enacted and experienced, either coming and going in very brief timeframes or existing as a background hum. As Pile (2010) explains, the underlying problem here is one of sequencing, and the inevitable involvement of the subjects' and researcher's own cognitive judgements during their conscious observations, reflections and articulations. Indeed any individual might witness, experience or add to a happening, but as soon as they reflect on it, they involve their personally, culturally and historically anchored thoughts and emotions which evoke a re-created consciousness of the happening – now one step removed from it in its pure original form (Conradson and Latham, 2007; Ducey, 2007). In response to both of these general challenges a partial mitigating strategy involves one or both of two approaches. First, if one has to write about happening or ask subjects to verbalize happenings, to describe them in detail and quite expressively in an 'irrealis mood', including the sensations involved, and be as true and honest as possible to their energy and momentum (Cadman, 2009; Ducey, 2007; Laurier and Philo, 2006; Vannini, 2014b). Second, to introduce complementary approaches that provide either secondary sensory insights or a new event themselves (such as photography, music and other art). In short, all this might be done so that legitimate attempts are made at 'relaying,' 'presenting' and 'making' realities more than (re)representing them, even though one can only mitigate so far. Indeed, in sum, and as Andrews (2014a) posits, characterizing non-representational theory is its inherent methodological flexibility, experimentation and playfulness that exists in three related forms: (1) a willingness to create, and produce, case-specific hybrid methodologies so that observations stay focused on the immediate and do not go too far down contemplative and interpretative paths (Patchett, 2010); (2) a willingness to embrace methods that move beyond forms of linguistic expression in both data capture and knowledge translation (Patchett, 2010); and (3) a willingness to 'montage' methods, where they are often juxtaposed and inhabit different space-times (see Latham, 2003), not to 'triangulate' in a purist and traditional methodological, but to themselves reflect the multiplicity and relationality of the world.

The current chapter notes the above methodological objectives, facets and challenges, and also a very valid observation, made by Cresswell (2012), that it is ironic that despite all the talk of the methodological radicalness and liveliness of NRT, it remains in it most common form, conventionally written; very 'texty'. Thus the chapter goes some way to address this and showcase alternatives. Indeed, acting into life – and building, changing and boosting it – by creating two musical moments that are clearly 'out there' for public consumption and experience. Although we unavoidably had to write this chapter, a YouTube video and Soundcloud MP3 of the songs give our words some additional life. Although there has been an emerging interest in the health and well-being aspects and implications of music in geography (Andrews, et al., 2011; 2014b) – including more-than-representational perspectives (Anderson, 2002; 2006; Andrews, 2014b; 2014c; Evans, 2014; McCormack, 2003; 2013; Simpson, 2014) – to our knowledge no research has purposefully created music or involved an artists' existing creations.

'The pump'

Figure 12.1 'The pump'

https://www.youtube.com/watch?v=nGjEn3iH4cI
http://www.shardcore.org/music/clubsofa/02%20the%20pump.mp3
(or search 'shardcore clubsofa the pump')

It's 1998 and, bar a paper or two, NRT is merely a glint in Nigel Thrift's eye. Newly out of his PhD program and short post-doc at the University of Nottingham, Gavin is just starting his first lectureship position at Buckinghamshire Chilterns University College (now Buckinghamshire New University), his early work in health geography taking a rather conventional approach to the political economy of care. Eric (also known in art-circles as Shardcore) had long since abandoned his PhD studies in cognitive psychology at Oxford, and was enjoying his fame as a character on MTV's *Real World* whilst being a founding director of Beenz.com, the world's first online currency with serious aspirations. We had known each other for over 15 years even back then, having grown up together in the small Devon seaside resort of Teignmouth (see Andrews and Kearns, 2005). Now, along with two other good friends – Matt, a PhD-prepared medical sociologist who was moving into health PR and Chris, a banker – we converge on London's East End (Stepney/Whitechapel) to live together and start our post-education lives, lives full of work but also full of experiencing London, parties, drinking, smoking and eating roast dinners. It is in this particular context and moment that 'The Pump' emerges quite organically, with an aim to make a comment and simultaneously put something fun and lively into the world. Eric, always the least sporty of the group, had suddenly found himself living with three fitness enthusiasts. Gavin was in the tail end of his childhood athletics (400m) hobby; Matt frequently ran, swam and played ball sports; and Chris had newly been introduced to the local

YMCA gym by Matt and Gavin and, very unlike his old self, had started asking people to 'feel his tris'.

Eric, at the time, happened to be in the middle of producing his first experimental electronic album *Club Sofa*, which itself took two years to complete. (The title was initially a phrase Chris used – an ironic statement on being too lazy or stoned to leave the house to go clubbing on a Friday night, yet the musical content providing one with a relaxed and trippy atmosphere at home in the club's place.) He decided to write a track for *Club Sofa* which was to comment on the fitness situation unravelling around him and body culture more generally. Indeed, for the gym skeptic Eric, the notion that fitness could provide the same kind of high as the more traditional routes of sex, drugs and rock and roll was an alien notion, whilst the associated homoeroticism of heterosexual men admiring each other's toned bodies was equally amusing (and at odds with the 'straight' demeanor presented particularly by Matt and Chris). Hence, 'The Pump' is a social commentary about the narcissism, homoeroticism, campiness and desire in training in gyms and men's body and fitness culture more generally. It is also motivated by a fascination for 'the other' and 'the freak', but at the same time, rather than offend, aims to be funny and entertaining. In retrospect, one could consider 'The Pump' to have a party political motivation. However, back in 1998 when the track was written, Arnold Schwarzenegger was a movie star, had not entered politics, and the term 'the governator' had not entered the popular vernacular. Even if he had been in party politics however, 'The Pump' would not necessarily have been made by Eric to create controversy or embarrassment because Arnold has always celebrated the film and the part of his life it conveys which launched his global fame. Indeed, he often attends reunion events pertaining to it, and speaks with real affection about it.

Gavin and Matt generally supported Eric's production of 'The Pump'. They were involved in the creative process to the extent of providing comments on the sounds and samples that gradually emerged from Eric's home office on evenings and weekends (largely pertaining to accuracy). On one level they thought the song was funny and thus acknowledged the silliness of their own gym obsession. On another level, however, it fit their own academic transitions and emerging interest in sports, fitness bodies and society (that was later realized in a number of publications – see Andrews and Andrews, 2003; Andrews, et al., 2005; Sudwell, 1999), and thus they appreciated the song as a critical academic intervention in itself. Of course, back then their academic interest was more aligned with the meaning of the social commentary in 'The Pump' (although regardless of what paradigm they were working within, its 'affects' were being created).

'The Pump', like much of Eric's music and the entirety of *Club Sofa*, might be classified as broadly as 'experimental electronic', the track perhaps being more specifically 'high energy industrial techno'. It was produced, mixed and stored by Eric on a basic 486 PC running Windows 95 with early Cakewalk software, a soundboard and Sound Blaster interface, and the first ever Roland TB 303 emulator for PCs. The song starts with the sound of a weight machine in action, the screech of metal scraping against metal in a regular rhythm (in reality a sample of a squeaky metal door opening, checked by Matt that it sounded like a weight machine as Eric had never actually been into a gym). The sound of Arnold Schwarzenegger's voice then quickly enters the fray (sampled from his 1977 docudrama *Pumping Iron* – a well-worn copy of which often frequented the household video player, providing perfect audio referents for Eric to lift and manipulate): 'The greatest feeling that you can get in the gym, or the most satisfying feeling is the pump.' Then enters the break beat (at 120 bpm), an off-beat sample and bass line. After a four bar drum sequence follows a quick roll-kick, and back in comes Arnold explaining, 'I'm like getting the feeling of cuming in the gym, I'm getting the feeling of cuming at home, I'm getting the feeling of cuming backstage when I pump up, when I pose out in front of five-thousand people. It feels fantastic!' After another drum roll enters a regular sample of a straining male grunt (actually a sample from the early nineties sci-fi 'shoot-em-up' video game Doom), dispersed lyrical returns to 'it feels fantastic', and 'so I am cuming day and night!' and the repeated signature refrain 'the pump'. Then, half way through, in come the overdriven 303s giving the song more of a recognizable nineties liquid 'acid' sound, whilst the lyrics repeat, seemingly in no particular order, apart from 'the pump . . . the pump . . . the pump . . . the pump'. Towards the end of the track, back in come the 303s now overdriven to the max, and emitting more of an industrial buzz. Finally, there are eight 'the pumps' and a lone musicless lyric that fades the song out: 'It's as satisfying to me as cuming is, you know, as having sex and cuming!'

When it's all over in 2 minutes and 24 seconds, the listener realizes that 'The Pump' did not really 'go anywhere', yet it still energized. This might be attributed to two factors. First, although not relying on intensity 'builds' or 'peaks and troughs' (common in much dance music), it generates positive energy through its ongoing and unresolved harmonic tension. The song never settles to a 'home chord', which sets up an underlying anticipation that is never released – only quickly recycled. Second, the song never reverts to a familiar verse-chorus framework. Instead it moves along loosely without any such road map, yet with constant power.

In terms of circulation, 'The Pump' was given away free online, as part of the CD *Club Sofa*, during the late 1990s, long before MP3 structures had been introduced or YouTube had arrived (a point that Eric is particularly proud of). Otherwise many physical copies were circulated amongst Eric's friends (both amounting to a few hundred copies). In 2005, after someone reminded him about the song, Eric posted 'The Pump' on YouTube with an accompanying video made from extracts from *Pumping Iron* and closely synchronized. By 2015 this video had attained almost 11,000 views. Meanwhile in 2008 the entire *Club Sofa* album, including 'The Pump', was made available on Eric's personal website as an MP3 download. At an academic meeting in 2014 we agreed that, insofar as 'The Pump' is concerned, both its affects and meanings have spread thinly and globally, but that they also have stood the test of time, remaining as an audio reminder of our past lives, what was going on and the feels involved.

'Senescence'

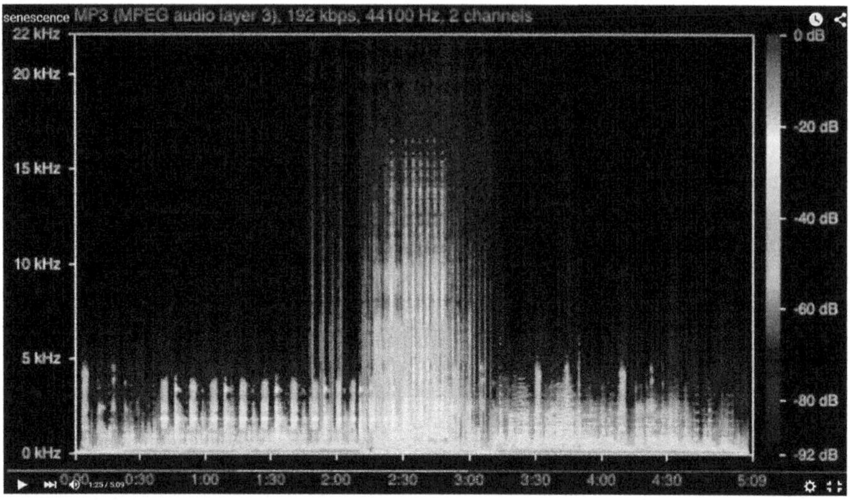

Figure 12.2 'Senescence'

https://www.youtube.com/watch?v=PVirqVH0Myw
http://www.shardcore.org/shardpress/index.php/2012/11/26/senescence/
(or search 'shardcore senescence')

It is 2012. NRT is no longer the new game in town; it's mainstream. Eric and Gavin are both married with kids, Eric living in Brighton, UK, and Gavin for over a decade in Ontario, Canada. Eric is now a full-time multimedia artist with a growing national reputation (occasionally dabbling academically in language acquisition and otherwise in IT when financial needs necessitate), whilst Gavin has moved through the academic ranks, chairing a new academic department along the way. His sabbatical at the end of this administrative role give him the time and space to develop his interest and knowledge in NRT. The motivation and underlying reason for the production of the song 'Senescence' is squarely academic and set within an appreciation of non-representational theory, although the context of the project changed. Gavin was putting together an edited collection entitled 'Soundscapes of Wellbeing in Popular Music' (Andrews, et al., 2014b), and Eric was part of the original authorship team. However, differences in expectations with other team members as to what Eric's authorship involved led to his withdrawal from that project. By the time this happened 'Senescence' had already been written, but luckily the opportunity soon arose to instead discuss it in the current chapter in this book. The change in publication venue, and the politics involved, illustrates how difficult it can be sometimes for artists and academics to work together, particularly given their very different worlds, ways and styles of working and end products.

The academic brief from Gavin was for Eric to produce a track in the style of genre of his choice that would simply comment on health or ageing and induce an affective sense of well-being, to make a more-than-representational intervention in health geography or geographical gerontology, directly acting into, building and boosting life. Beyond this Eric had a clean slate, and Gavin really did not know what would be produced (perhaps in retrospect, knowing Eric, expecting direct political commentary on global health, health information, the corporatization in health care or the state of the British NHS). On visiting Eric in early 2013, what Gavin heard was completely unexpected.

The track is slow, ambient and electronic, opening with a first phase (0:00–1:15) composed of the breathing in and out of a body, a heartbeat beat, a haunting piano playing sparingly in the background and regular bodily gurgles. After a lead-in (1:50–2:10), the second phase (2:10–2:50) increases in tempo notably. Metallic ticking sounds replace the heartbeat beat, and stressful in-time grunts and other percussive samples emerge. This second phase is faster and more frantic, but is soon over. After another short lead-in (2:50–3:20), a third phase (3:20–5:09) then slows everything down once again – but even slower than the first – acting like a prolonged fade-out. The heartbeat beat returns but is accompanied with complex fracturing of rhythm and broken piano melody. The song, reflecting its title, is a midlife audit made purely from samples of Eric's own body with an artificial sub-base. The three musical phases reflect three phases of a body's life: (1) normal, youthful stress–free, with optimal cell replication; (2) midlife, hard work with consequences in terms of stress on the body and mind; and (3) later life, less stress with the replication of cells broken as the body breaks down.

The idea for a slow, literally organic, yet electronic song was Eric's first. When he was asked by Gavin to think about producing the track, he was precisely at the midpoint of his own life (as projected by the UK Office of National Statistics). Moreover, like most people of his age, he realized midlife was upon him because he had started noticing his own body breaking down. Thus Eric wondered whether this could be captured in some way in a song, alongside a scientific understanding. In other words, whether he could conveying both the physical effects of aging and the internal sense of fatigue and the reality that all of these effects are driven by cell senescence – their breakdown in replication at a microscopic physical scale. In terms of artistic decisions, shortly after receiving Gavin's request, Eric thought to himself, 'If a chapter could be a noise, what noise would it make?' His skills in technology and electronic music created the possibility of reproducing the body in music through sampling its sounds. Moreover, Eric reflects in hindsight that the song was never going to be a pop song with lyrics or any other genre. (Did he never think, for example, about creating a happy sound to produce some happy affect?) As Eric reflects, 'The song has affect anyway, and it reflected me as a middle aged man. It was not for a crowd of hungry teenagers.'

Although stripped down, 'Senescence' benefitted from the latest technology, which was far more advanced than what was available for 'The Pump' 14 years earlier. In terms of hardware, it was created on a Powermac desktop with an Oxygen M8 Mini Keyboard controller (although a homemade stethoscope and microphone interfaced with Eric's body, providing access to his bodily sounds). In terms of software, it was sequenced in GarageBand and recorded using Audacity. Recording and production took only 10 days, much of this time used finding the right noises. In terms of circulation, an MP3 of 'Senescence' has been posted on Eric's personal and professional website since November 2012. To date it has been downloaded almost 300 times, mainly by artists and students broadly interested in Eric's work (that also involves painting, algorithmic art and installation work). Eric's website includes a personal, public and academic narrative about the song and well-being much more broadly, complete with bright illustrations. Mirroring the song, this information is provocative rather than fully explanatory, leading readers just to the door of diverse literatures and debates, including cell biology, music and time, emotion and soul, and rhythm in humans and human social history (see Figure 12.3). In April 2015, to coincide with the production of the current chapter, the track was also made available on YouTube.

Figure 12.3 'Senescence' and its narrative on Eric's website

Conclusion

Arts are both an emerging empirical interest and expanding form of methodological practice in health geography, including for example in the realms of fictional literature (Gesler, 2000; Baer and Gesler, 2004; Tonnellier and Curtis, 2005; Williams, 2007), painting and sculpture (Evans, et al., 2009; Parr, 2006; 2007; McCabe, et al., 2007) and musical sound, dance and movement (Anderson, 2002; 2006; Andrews, 2014b; 2014c; Atkinson and Rubidge, 2013; Evans, 2014; McCormack, 2003; 2013; Simpson, 2014). Through the creation of art (music), this chapter illustrates an innovative form of engagement for artists and academics to re-represent the world of health through its pure physicality by introducing a non-representational less-than-conscious physical change, a basic energy and movement into the world. Indeed, the two musical tracks clearly evoke reflection about the meaning of health. Yet, as musical patterns and rhythms, they also, more fundamentally, produce aligned affects ('The Pump' being youthful and energetic, whilst 'Senescence' is slower and deconstructed). These two facets – meaning and affect – are not always specifically isolated by the artist in the production of music, although a degree of ascetic choice, of course, exists, in terms of what an artist is attempting to do at any one time (and the two are heavily interconnected). In terms of circulation, both meanings and affect spread together as the music is downloaded and passed on, the 'research' acting into the world and building new physical and attitudinal realities, even if within each individual 'consumer' these are modest. This, we think, is what Thrift had in mind methodologically for a new paradigm of research in human geography based on NRT: a new politics underpinned by actual performance and movement (Thrift, 2008). This is what we have tried to achieve at some modest level.

The two musical tracks showcase two ways that music and academic health geography might emerge together under a non-representational tradition, through collaboration between artists and researchers either accidentally and organically (through mutual association) or purposefully through design. If NRT is to become even more commonplace across health geography, art would play an important part, and thus these are the types of collaborations that will have to occur, lest it becomes, as Cresswell (2012) fears, a 'texty' – not to mention ironic – theoretical exercise. Indeed this is somewhat new territory for health geography, and challenges exist. As we suggested at the start of this chapter, significant challenges arise when different types of professional work together and cross mediums (a researcher with and artist and their sounds, and an artist with an academic and their academic words). Fitting music into scholarship and scholarship into music is not easy, and both parties must endeavor to understand one another's motivations, priorities, expectations and methods. In future it will be necessary to move beyond the 'trial and error' approach – as we used in these initial reconnaissance's – and locate work within the literature on arts in academia, and specifically on artistic practice as research and knowledge translation (e.g. Barrett and Bolt, 2014; Dewey, 2007; Haseman, 2006; Kontos and Poland, 2009; Lapum, et al., 2012; Rossiter, et al., 2008; Sinding, et al., 2011). This engagement is important, not only for maximizing the success and outcomes of individual projects, but in producing general progress and a strong overall body of work, particularly in the face of institutional obstacles such as expectations of research output (particularly in career contexts such as Research Excellence Framework audits, tenure and promotion and so on).

The authors are still trying to figure one another out on a professional level, but hopefully have established a sound basis for future collaborations. To date our collaborations have involved a sharp division of labor, Eric in charge of the soundscape and Gavin in charge of the textscape. In future we might mix these roles a little more so that our projects become a little less multidisciplinary and a little more transdisciplinary in orientation. We continue to note and take inspiration from comments by Anderson and Harrison (2010), who ask, if culture comes so easy to all of us, why does it come so hard to academics who have tended to overthink it? Why when the postmodern turn in popular culture has involved so much spectacle, energy and action, has it involved so much serious, static and heavy theorization by academics? We do not know the answers to these questions, but we do know that despite the aforementioned challenges in our collaboration, doing NRT in health geography did not come so hard to us, was not overthought by us and was full of life.

References

Anderson, B., 2002. A principle of hope: recorded music, listening practices and the immanence of utopia. *Geografiska Annaler: Series B, Human Geography*, 84(3–4), pp. 211–227.

Anderson, B., 2006. Becoming and being hopeful: towards a theory of affect. *Environment and Planning D*, 24(5), pp. 733–752.

Anderson, B., and Harrison, P., 2010. *Taking Place: non-representational theories and geography*. Surrey: Ashgate.

Andrews, G. J., 2014a. Co-creating health's lively, moving frontiers: brief observations on the facets and possibilities of non-representational theory. *Health and Place*, 30, pp. 165–170.

Andrews, G. J., 2014b. 'Gonna live forever': Noel Gallagher's spaces of wellbeing. In Andrews, G. J., Kingsbury, P., Kearns, R. A. *Soundscapes of and wellbeing in popular music*. Ashgate, London.

Andrews, G. J., 2014c. A force from the beginning: wellbeing in the affective intensities of pop music *Aporia* 6(4), pp. 6–18.

Andrews, G. J., 2015. The lively challenges and opportunities of non-representational theory: a reply to Hanlon and Kearns. *Social Science and Medicine*, 128, pp. 338–341.

Andrews, G. J., Chen, S., and Myers, S., 2014a. The 'taking place' of health and wellbeing: towards non-representational theory. *Social Science and Medicine*, 108, pp. 210–222.

Andrews, G. J., Evans, J., and McAlister, S., 2013. 'Creating the right therapy vibe': Relational performances in holistic medicine. *Social Science and Medicine*. 83, pp. 99–109.

Andrews, G. J., and Kearns, R. A., 2005. Everyday health histories and the making of place: the case of an English coastal town. *Social Science and Medicine*, 60(12), pp. 2697–2713.

Andrews, G. J., Kearns, R. A., Kingsbury, P., and Carr, E. R., 2011. Cool aid? Health, wellbeing and place in the work of Bono and U2. *Health and Place*, 17(1), pp. 185–194.

Andrews, G. J, Kingsbury P., Kearns, R. A., 2014b. *Soundscapes of and Wellbeing in Popular Music*. London: Ashgate.

Andrews, G. J, and Shaw, D., 2010. 'So we started talking about a beach in Barbados': visualization practices and needle phobia. *Social Science and Medicine* 71, pp. 1804–1810.

Andrews, G. J., Sudwell, M. I., and Sparkes, A. C., 2005. Towards a geography of fitness: an ethnographic case study of the gym in British bodybuilding culture. *Social Science and Medicine*, 60(4), pp. 877–891.

Andrews, J. P., and Andrews, G. J., 2003. Life in a secure unit: the rehabilitation of young people through the use of sport. *Social Science and Medicine*, 56(3), pp. 531–550.

Atkinson, S., and Rubidge, T., 2013. Managing the spatialities of arts-based practices with school children: an inter-disciplinary exploration of engagement, movement and well-being. *Arts and Health*, 5(1), pp. 39–50.

Atkinson, S., and Scott, K., 2015. Stable and destabilised states of subjective well-being: dance and movement as catalysts of transition. *Social and Cultural Geography*, 16(1), pp. 75–94.

Baer, L. D., and Gesler, W. M., 2004. Reconsidering the concept of therapeutic landscapes in JD Salinger's *The Catcher in the Rye*. *Area*, 36(4), pp. 404–413.

Barnfield, A., 2015. Public health, physical exercise and non-representational theory – a mixed method study of recreational running in Sofia, Bulgaria. *Critical Public Health*, (forthcoming).

Barrett, E., and Bolt, B. (eds.), 2014. *Practice as research: approaches to creative arts enquiry*. London: Ib Tauris.

Bissell, D., 2010. Placing affective relations: uncertain geographies of pain. In Anderson, B., Harrison, P. (eds.) *Taking place: non-representational theories and geography*. Aldershot: Ashgate, pp. 79–98.

Boyd, C., and Duffy, M., 2012. Sonic geographies of shifting bodies. *Interference: A Journal of Audio Culture*, 1(2), pp. 1–7.

Boyer, K., 2012. Affect, corporeality and the limits of belonging: breastfeeding in public in the contemporary UK. *Health and Place*, 18(3), pp. 552–560.

Braun, B., 2007. Biopolitics and the molecularization of life. *Cultural Geographies*, 14(1), pp. 6–28.

Cadman L., 2009. *Nonrepresentational theory / nonrepresentational geographies*. In Kitchin, R., and Thrift, N. (eds.) *International Encyclopedia of Human Geography*. Philadelphia: Elsevier. p. 2–9.

Conradson, D., and Latham, A., 2007. The affective possibilities of London: antipodean transnationals and the overseas experience. *Mobilities*, 2(2), pp. 231–254.

Cresswell, T., 2012. Nonrepresentational theory and me: notes of an interested sceptic. *Environment and Planning D: Society and Space*, 30(1), pp. 96–105.

Dewey, J., 2007. The marriage of art and academia: challenges and opportunities for music research in practice-based environments. *Dutch Journal of Music Theory*, 12(1), pp. 34–40.

Dewsbury, J D., 2003. Witnessing space: knowledge without contemplation. *Environment and Planning A*, 35, pp. 1907–1932.

Dewsbury, J. D., 2009. *Performative, non-representational and affect-based research: Seven injunctions, The Sage Handbook of Qualitative Geography*. London: Sage.

Dewsbury, J. D., Harrison, P., Rose, M., and Wylie, J., 2002. Enacting geographies. *Geoforum*, 33(4), pp. 437–440.

Dirksmeier, P., and Helbrecht, I., 2008. Time, non-representational theory and the 'performative turn' – towards a new methodology in qualitative social research. In *Forum: Qualitative Social Research*, 9, p. 2.

Ducey, A., 2007. More than a job: meaning, affect, and training health care workers. In Clough, P., and Halley, J. (eds.) *The affective turn: theorizing the social*. Durham, NC: Duke University Press. pp. 187–208.

Duff, C., 2010. Towards a developmental ethology: exploring Deleuze's contribution to the study of health and human development. *Health*, 14(6), pp. 619–634.

Duff, C., 2011. Networks, resources and agencies: on the character and production of enabling places. *Health and Place*, 17(1), pp. 149–156.

Duff, C., 2012. Exploring the role of 'enabling places' in promoting recovery from mental illness: a qualitative test of a relational model. *Health and Place*, 18(6), pp. 1388–1395.

Duff, C., 2014. *Assemblages of Health*. New York: Springer.

Evans, J. D., 2014. Painting therapeutic landscapes with sound: *On Land* by Brian Eno. In Andrews, G. J., Kingsbury, P., Kearns, R. A. (eds.), 2014. *Soundscapes of and wellbeing in popular music*. London: Ashgate.

Evans, J. D., Crooks, V. A., and Kingsbury, P. T., 2009. Theoretical injections: on the therapeutic aesthetics of medical spaces. *Social Science and Medicine*, 69(5), pp. 716–721.

Foley, R., 2011. Performing health in place: the holy well as a therapeutic assemblage. *Health and Place*, 17(2), pp. 470–479.

Foley, R., 2014. The Roman-Irish bath: medical/health history as therapeutic assemblage. *Social Science and Medicine*, 106, pp. 10–19.

Gatrell, A. C., 2013. Therapeutic mobilities: walking and 'steps' to wellbeing and health. *Health and Place*, 22, pp. 98–106.

Gesler, W., 2000. Hans Castorp's journey-to-knowledge of disease and health in Thomas Mann's *The magic mountain. Health and Place*, 6(2), pp. 125–134.

Greenhough, B., 2006. 'Decontextualised? Dissociated? Detached? Mapping the networks of bio-informatic exchange' *Environment and Planning A*, 38(3), pp. 445–463.

Greenhough, B., 2011a. Assembling an island laboratory. *Area*, 43(2), pp. 134–138.

Greenhough, B., 2011b. Citizenship, care and companionship: approaching geographies of health and bioscience. *Progress in Human Geography*, 35(2), pp. 153–171.

Greenhough, B., and Roe, E. J., 2011. Ethics, space, and somatic sensibilities: comparing relationships between scientific researchers and their human and animal experimental subjects. *Environment and Planning D: Society and Space*, 29(1), pp. 47–66.

Hall, E., 2004. Spaces and networks of genetic knowledge making: the 'geneticisation' of heart disease. *Health and Place*, 10(4), pp. 311–318.

Hanlon, N., 2014. Commentary: doing health geography with feeling. *Social Science and Medicine*, 115, pp. 144–146.

Haseman, B., 2006. A manifesto for performative research. *Media International Australia, Incorporating Culture and Policy*, 118, (February): pp. 98–106. Available at: http://search.informit.com.au/documentSummary;dn=010497030622521;res=IELLCC> ISSN: 1329-878X.

Jackson, P., and Neely, A. H., 2014. Triangulating health: toward a practice of a political ecology of health. *Progress in Human Geography*, 39(1), pp. 47–64.

Kearns, R. A., 2014. The health in 'life's infinite doings': a response to Andrews et al. *Social Science and Medicine*, 115, pp. 147–149.

Kearns, R., and Moon, G., 2002. From medical to health geography: novelty, place and theory after a decade of change. *Progress in Human Geography*, 26(5), pp. 605–625.

Kontos, P. C., and Poland, B. D., 2009. Mapping new theoretical and methodological terrain for knowledge translation: contributions from critical realism and the arts. *Implementation Science*, 4(1), pp. 1–10.

Kraftl, P., and Horton, J., 2007. 'The health event': everyday, affective politics of participation. *Geoforum*, 38(5), pp. 1012–1027.

Lapum, J., Ruttonsha, P., Church, K., Yau, T., and David, A. M., 2012. Employing the arts in research as an analytical tool and dissemination method interpreting experience through the aesthetic. *Qualitative Inquiry*, 18(1), pp. 100–115.

Latham, A., 2003. Research, performance, and doing human geography: some reflections on the diary-photograph, diary-interview method. *Environment and Planning A*, 35(11), pp. 1993–2018.

Latham, A., and Conradson, D., 2003. The possibilities of performance. *Environment and Planning A*, 35(11), pp. 1901–1906.

Laurier, E., 2005. Searching for a parking space. *Intellectica*, 2–3(41–42), pp. 101–116.

Laurier, E., and Philo, C., 2006. Possible geographies: a passing encounter in a café. *Area*, 38(4), pp. 353–363.

Lea, J., Cadman, L., and Philo, C., 2014. Changing the habits of a lifetime? Mindfulness meditation and habitual geographies. *Cultural Geographies*, 22(1), pp. 49–65.

Lorimer, H., 2005. Cultural geography: the busyness of being 'more-than-representational'. *Progress in Human Geography*, 29(1), pp. 83–94.

Lorimer, H., 2006. Herding memories of humans and animals. *Environment and Planning D: Society and Space*, 24, pp. 497–518.

Lorimer, H., 2007. Cultural geography: worldly shapes, differently arranged. *Progress in Human Geography*, 31(1), p. 89.

Lorimer, H., 2008. Cultural geography: nonrepresentational conditions and concerns. *Progress in Human Geography*, 32(4), pp. 551–559.

Macpherson, H., 2008. 'I don't know why they call it the Lake District they might as well call it the rock district!' The workings of humour and laughter in research with members of visually impaired walking groups. *Environment and planning. D, Society and space*, 26(6), p. 1080.

Macpherson, H., 2009a. The intercorporeal emergence of landscape: negotiating sight, blindness, and ideas of landscape in the British countryside. *Environment and Planning A*, 41(5), p. 1042.

Macpherson, H., 2009b. Articulating blind touch: thinking through the feet. *The Senses and Society*, 4(2), pp. 179–193.

McCabe, K., Summerton, R., and Parr, H., 2007. Mental health via cultural citizenship. *Journal of Public Mental Health*, 6(4), pp. 33–36.

McCormack, D. P., 2003. An event of geographical ethics in spaces of affect. *Transactions of the Institute of British Geographers*, 28(4), pp. 488–507.

McCormack, D. P., 2013. *Refrains for moving bodies: experience and experiment in affective spaces*. Durnham, NC: Duke University Press.

Parr, H., 2006. Mental health, the arts and belongings. *Transactions of the Institute of British Geographers*, 31(2), pp. 150–166.

Parr, H., 2007. Collaborative film-making as process, method and text in mental health research. *Cultural Geographies*, 14(1), pp. 114–138.

Patchett, M., 2010. *Putting animals on display: geographies of taxidermy practice*. PhD thesis, University of Glasgow.

Paterson, M., 2005. Affecting touch: towards a felt phenomenology of therapeutic touch. In Davidson, J., Bondi, L., and Smith, M. (eds.), *Emotional geographies*. Aldershot: Ashgate. pp. 161–173.

Philo, C., 2007. A vitally human medical geography? Introducing Georges Canguilhem to geographers. *New Zealand Geographer*, 63(2), pp. 82–96.

Philo, C., 2012. A 'new Foucault' with lively implications – or 'the crawfish advances sideways'. *Transactions of the Institute of British Geographers*, 37(4), pp. 496–514.

Philo, C., Cadman, L., Lea, J., 2014. New energy geographies: a case study of yoga, meditation and healthfulness. *Medical Humanities*, 36(1), pp. 35–46.

Pile, S., 2010. Emotions and affect in recent human geography. *Transactions of the Institute of British Geographers*, 35(1), pp. 5–20.

Rossiter, K., Kontos, P., Colantonio, A., Gilbert, J., Gray, J., and Keightley, M., 2008. Staging data: theatre as a tool for analysis and knowledge transfer in health research. *Social Science and Medicine*, 66(1), pp. 130–146

Simpson, P., 2014. A soundtrack to the everyday: street music and the production of convivial, healthy public spaces. In Andrews, G. J., Kingsbury, P., Kearns, R. A. (eds.) *Soundscapes of and wellbeing in popular music*. London: Ashgate. pp. 159–173.

Sinding, C., Schwartz, L., and Hunt, M., 2011. Staging ethics: the promise and perils of research-based performance. *Canadian Theatre Review*, 146(1), pp. 32–37.

Solomon, H., 2011. Affective journeys: the emotional structuring of medical tourism in India. *Anthropology and Medicine*, 18(1), pp. 105–118.

Sudwell, M., 1999. The body bridge. *Talking bodies: Men's narratives of the body and sport*. Jyväskylä, Finland: SoPhi Academic Press. pp. 13–28.

Thrift, N., 1997. The still point: resistance, expressive embodiment and dance. In Pile, S., and Keith, M. (eds.) *Geographies of Resistance*, London: Routledge. pp. 124–151.

Thrift, N., 2000. Afterwords. *Environment and Planning D-Society and Space*, 18(2), pp. 213–255.

Thrift N., 2004. Intensities of feeling: towards a spatial politics of affect. *Geografiska Annaler B*, 86, pp. 57–78.

Thrift, N., 2008. *Non-representational theory: space, politics, affect*. London: Routledge.

Timmons, S., Crosbie, B., and Harrison-Paul, R., 2010. Displacement of death in public space by lay people using the automated external defibrillator. *Health and Place*, 16(2), pp. 365–370.

Tonnellier, F., and Curtis, S., 2005. Medicine, landscapes, symbols: *The country doctor* by Honoré de Balzac. *Health and Place*, 11(4), pp. 313–321.

Tucker, I., 2010. Everyday spaces of mental distress: the spatial habituation of home. *Environment and Planning. D, Society and Space*, 28(3), p. 526.

Vannini, P., 2009. Nonrepresentational theory and symbolic interactionism: shared perspectives and missed articulations. *Symbolic Interaction*, 32(3), pp. 282–286.

Vannini P., 2014a. Non-representational research methodologies: an introduction. In Vaninni, P. (ed.) *Non-representational methodologies: re-envisioning research*. London: Routledge.

Vannini, P., 2014b. Enlivening ethnography through the irrealis mood: in search of a more-than-representational style. In Vaninni P (eds.) *Non-representational methodologies: re-envisioning research*. London: Routledge.

Williams, A., 2007. Healing landscapes in the Alps: *Heidi* by Johanna Spyri. In Williams, A. *Therapeutic Landscapes*. Aldershot: Ashgate. pp. 65–77.

Wylie, J., 2005. A single day's walking: narrating self and landscape on the South West Coast Path. *Transactions of the Institute of British Geographers*, 30(2), pp. 234–247.

13 Managing and overcoming the challenges of qualitative research on palliative family caregivers

Allison Williams

Reflections on praxis

Based on an ongoing research project with palliative family caregivers, this chapter operates as a reflection on praxis, to identify ethical, practical and emotional issues that permeate this type of often sensitive research given the critical nature of the population of interest. Discussion specific to how these issues can be managed in order to better facilitate research, which will ultimately enhance the contributions made to health care policy and practice, is at the heart of this chapter. Although the logistic and emotional challenges for the researcher are many, given the clear need for often fragile caregiver participants to communicate with a distant, nonjudgmental listener, the rewards have been great. Not only is speaking to individual family caregivers (FCGs) rewarding, in that they personify compassion, but speaking in front of Parliament in Ottawa, Canada's capital city, confirms that the research knowledge is making a difference in the policy and program arenas.

Introduction

Qualitative research methods are a distinguishing feature of social scientific research, enabling social scientists to construct accounts of why place matters to health and health care across a wide range of issues (Dyck, 1999; Kearns and Moon, 2002). This includes attention to various aspects of formal and informal care and caregiving, such as the impact of health care reform on informal FCGs (Milligan, et al., 2007). This work has challenged assumptions and altered perceptions about the people and places involved in care and caregiving (Donovan and Williams, 2011; Williams and Donovan, 2007; Williams, 2004; Brown, 2003; Dyck, 2001; Dyck, et al., 2005; Milligan, 2003; Wiles, 2003).

Qualitative research is recognized as a 'social process' (Dowling, 2005, p. 26) because it relies on human interaction as a means to understand and interpret various social phenomena. As such, issues of power and subjectivity permeate qualitative inquiry, with implications for the research participants, the researcher, and society as a whole. In research on family caregiving, these research issues are diverse and complex and complicated by the wide range of diverse situations in which people find themselves providing care. With the thoughtful guidance of research ethics boards (REBs), who have proven to facilitate the careful implementation of this work, qualitative researchers can address these issues over the course of their studies via the process of critical reflection (Dowling, 2005). In fact, reflexive processes help shape all aspects of qualitative research, including who and what we study; how we collect, analyze and interpret the data; and how we, as researchers, affect and are affected by the research. Such processes are known to increase the trustworthiness of the findings, an important element of rigor in qualitative research (Baxter and Eyles, 1997; Cutcliffe, 2003; Dowling, 2005).

As much as these processes clearly influence data collection and interpretation, readers are seldom privy to this 'inside knowledge,' given that published reports typically focus on research outcomes, rather than the challenges associated with the research process. Indeed, this is hardly surprising because the ultimate goal of the research program is to increase awareness about particular topics or issues. However, the failure to discuss research challenges can result in the replication of problematic choices and strategies in similar research endeavors. In this chapter, challenges experienced in a program of research spanning 15 years, specific to informal family caregiving during home-based palliative and end-of-life (P/EOL) care, are discussed. P/EOL care is defined by the World Health Organization (2007) as a philosophy of care which seeks to reduce the physical, emotional and spiritual suffering of patients and families who are facing a life-threatening or life-limiting illness. For the remainder of this chapter, this will be simply referred to as palliative care. The goal of the research program is to create awareness, and in so doing, inform potential solutions that will facilitate future research endeavors, and ultimately, enhance the quality of research outcomes which will potentially impact health policy and practice decisions.

Based on evidence collected from the experience of leading research involving palliative family caregivers, research challenges have been synthesized into three substantive areas: (1) ethical issues, (2) practical issues related to accessing and recruiting participants, and (3) emotion work required of the researcher. While presented individually, the overlap between these three substantive issues and their impact on the research process will be revealed. The chapter begins with a brief overview of palliative family caregivers. Following this introduction, more detail is provided on the above-mentioned substantive issues. The chapter concludes with a discussion regarding the implications specific to current and future qualitative palliative care research. This work is relevant to both new and experienced qualitative researchers, including those who are engaged in similarly sensitive qualitative and palliative care research.

Palliative family caregivers

Although FCGs are integral to the sustainability of the health care system in Canada (Romanow, 2002), further research is required to inform the improvement and development of appropriate palliative care services to protect their health so that they can meet the needs of their dying loved-ones (Grande, et al., 2009).

It is suggested that qualitative approaches are particularly well-suited for palliative care research (Koenig, Back and Crawley, 2003). In fact, their use has increased substantially over the last decade and dominates current research in this area (Payne and Turner, 2008). Research suggests that both family members[1] and FCGs prefer qualitative interviews over quantitative survey instruments because of the opportunity to elaborate on their situations (Buckle, et al., 2010; Gysels, et al., 2008a; Sherman, et al., 2005). Qualitative research methods are especially useful for research that involves the evaluation and efficacy of services and interventions (Williams, et al., 2008), social and cultural differences (Williams, 2002; Grande, et al., 2009) and many types of meaning-centered questions which are not easily addressed using other research techniques (Koenig, et al., 2003).

Palliative care research is regarded as profoundly sensitive, given its focus on work with the very ill, dying, and their families (Addington-Hall, 2007; Johnson and Clarke, 2003). It is suggested that nowhere in health care research is this sensitivity more critical, given the scope and intensity of the often 'unanswerable' issues with which the dying, their families, and health care practitioners encounter over the course of care and bereavement (Crossley, 2007, p. 181). As such, palliative care research is infused with unique practical and ethical challenges – beyond those which characterize qualitative research in general (Addington-Hall, 2007; Grande, et al., 2009; Koenig, et al., 2003). These challenges – including recruitment difficulties, high attrition rates and small sample sizes – impact both the quality and quantity of palliative care research.

Research challenges and strategies

Ethical issues in palliative care research with family caregivers

In palliative care research, ethical issues abound particularly with respect to vulnerability and informed consent (Casarett and Karlawish, 2000; Duke and Bennett, 2010; Koenig, et al., 2003). Vulnerability refers to the 'increased potential that one's interests cannot be protected' (Agrawal, 2003, p. S26). This is often associated with palliative populations based on the assumption that these groups are in need of greater protection because they are in situations that may render them incapable of making informed decisions about participation. With palliative patients this is somewhat intuitive, given that their decision-making abilities might be compromised by their deteriorating health status (although this debate is questioned, see Casarett and Karlawish, 2000; Koenig, et al., 2003). However, the issue of vulnerability as it relates to FCGs is less clear. FCGs are often perceived by REBs as having this same sense of vulnerability as the patients they care for, requiring that consent be obtained cautiously. This challenge is not adequately

1 The terms 'patients', 'clients' and 'family members' are used interchangeably in the literature to refer to the person for whom the care is provided. Except when directly quoted from a specific source, I use the term 'family member'. This is because I speak from the perspective of FCGs, and they do not view this person as a patient or client.

addressed in the palliative care literature (Payne and Turner, 2008), perpetuating the perception that both palliative patients *and* FCGs are vulnerable and thereby needed to be treated with great sensitivity (Casarett and Karlawish, 2000; Duke and Bennett, 2010; Koenig, et al., 2003).

The issue of vulnerability has led to significant difficulties in obtaining university ethics clearance. This is reflective of the fact that university REBs, as seen in previous literature (Casarett and Karlawish, 2000; Duke and Bennett, 2010; Koenig, et al., 2003), tend to impose harsher restrictions for palliative care research. In my experience, REBs perceive the aforementioned issue of palliative patient vulnerability as translating into FCG vulnerability; this is no doubt because of the REBs obligation to ensure that those conducting research with vulnerable groups have the skills and sensitivity to do so. Certainly research has shown that the demands of caregiving, together with a lack of adequate and appropriate support, contribute to stress in FCGs that is unique from palliative patients receiving the care (Donovan and Williams, 2011; Williams and Donovan, 2007; Carstairs, 2010; Stajduhar and Cohen, 2009). Even so, the concern over vulnerability translates into REBs implementing a rigorous screening process on applications associated with palliative care – even if only for the FCGs; this often requires numerous amendments.

In particular, I have faced scrutiny by the REB because of our disciplinary background as social scientists and not health care researchers (i.e., not clinical health scientists who are in direct contact with this population and thereby better placed to do this research) and because of our focus on palliative FCGs, and not the family members for whom they care for. In fact, I have been interviewed face-to-face by the university REB on two separate occasions to elaborate on the relevance and potential contribution of a social science perspective in research on palliative family caregiving, as well as prove my ability to carry out the work in a sensitive manner. Furthermore, the REB questioned our proposed methods and recruitment processes. For example, there was concern that the longitudinal approach used in our research – such as the number of interviews (n=4 on average) and their timing during the stages of caregiving and bereavement to capture changes over time – was too arduous for FCGs. In addition, working with health care providers to recruit participants (as later discussed) was regarded as potentially coercive by the REB because FCGs might feel compelled to participate for fear of jeopardizing the services that they or the family member received. Furthermore, it was suggested that FCGs might be inclined to hide their true feelings so that they do not say anything that could offend their family members or their health care providers. The responses to these questions are addressed below. The responses reveal the REBs concern regarding a researcher's skill set specific to engaging with FCGs, such as having the appropriate communication skills that mitigate harm.

Managing ethical challenges

Although I recognize the difficulty in asking FCGs to participate in longitudinal research when they are facing a physically and emotionally stressful situation, I feel strongly about the fact that this research is necessary to understand how to best support FCGs in their capacity as both recipients and providers of palliative care. Furthermore, our experience has shown that FCGs' involvement in the research process has, in fact, been therapeutic for them. Many participants have told us that having a non-judgmental listener allowed them to talk freely, honestly and fully about their experience and by so doing, allowed them to work a few things out about why and how they are caregiving. To respond to the concerns raised by REBs, I implemented a number of strategies and safeguards to reduce the burden of participation and preserve the integrity of participants.

I have learned, thanks to the REB process, that it is extremely important to be very clear and intentional about providing complete and transparent informed consent, even if it means a lengthy, albeit detailed and comprehensive letter of information/consent forms (LOI/CFs). In fact, the average length of LOI/CFs typically used in the projects within this research program was five pages. These documents covered the purpose and procedures involved in the research, our experiences with the research, the risks and benefits of participation, the right to withdraw and the protection of confidentiality. Furthermore, it was clearly indicated that the researchers involved were not directly affiliated with any of the health care providers with whom the research team worked with and was strongly emphasized that FCGs' decisions about participation would in no way affect current or future service provision. To mitigate burden, the LOI/CFs were reviewed at the beginning of each interview, and FCG participants would be asked if they were willing to continue. To address the issue specific to service provider recruitment potentially impacting service provision, the LOI made clear that involvement in the research would not have any effect on formal service provision. Once the REB saw the noted changes to our protocol, they gained trust in my research team's ability to carry out the work sensitively and gave the green light to proceed. Addressing the REB concerns only strengthened the research protocols of concern herein.

A key component of the research approach was what the researchers termed a 'participant-driven' data collection strategy; this ensured that FCGs had complete flexibility and control in terms of the timing and location of data collection and participation. To illustrate, the scheduling, timing, and interview setting was all determined by FCGs. Most interviews were held in FCGs' homes; however, researchers also met participants in coffee shops, libraries, and places of employment, when so requested. Furthermore, FCGs had the right to decide whether interviews would be audio-recorded by the researchers, to refuse to answer questions, to review transcripts or summaries, and to change, add or delete information. Interestingly, very few took up these options. Furthermore, although the research project often proposed the use of several ethnographic data collection techniques (e.g. observations, photovoice or photography), FCGs could decline those of which they felt uncomfortable. Finally, FCGs were free to decide if they wanted the family member present (or not) during the interviews. Most FCGs preferred to be interviewed on their own to not only allow them to speak freely, honestly and fully about their experiences but also to not add burden to their very sick family member. One exception was a spousal

FCG who wished to have the family member present to make him feel 'a part' of the process and have him involved as much as possible.

Attention to the concerns noted above is necessary to design studies that are both rigorous and sensitive, while increasing the likelihood of recruiting and retaining FCG participants, as will be discussed below. Furthermore, building in this flexibility was intended to demonstrate respect toward the participants; offering these controls and then honoring FCG's decisions was integral to trust-building, an element of emotion work, as will be discussed below.

Practical issues related to accessing and recruiting participants

Despite concerted efforts, difficulties in recruiting FCG participants have been encountered. In fact, this is common in palliative care research (Addington-Hall, 2007) and is regarded as one of the most challenging methodological and ethical issues encountered in socially sensitive research (Williams, et al., 2008). I experienced systemic and participant-specific barriers that made it difficult to identify and access FCG participants, and I provide an overview of strategies on how these were overcome.

There are at least five systemic barriers that made recruiting participants difficult, especially when working to recruit via health care and social service organizations. First, the lack of palliative care services in general limited the pool of potential participants. Related to this is the issue of prognostication, which refers to the time of referral to the palliative care caseload. Family members are often referred very late in the disease trajectory, if at all, resulting in a very small window of time in which to access participants. Second, recruiting participants was difficult because of gatekeeping. Common in health care research (Duke and Bennett, 2010; Miller, et al., 2003; Steinhauser, et al., 2006), gatekeeping refers to the reluctance by health care providers to approach potential participants, often because of perceived vulnerability and their own discomfort or biases toward the research topic or process. Concerns specific to the latter included the belief that their organizations would not benefit from the research and that the focus would be on palliative FCGs and not family members. Third, the use of the term 'palliative' was uncomfortable for many health care providers, given that many FCGs and family members had not accepted such a prognosis. Many identified FCGs were not approached about the research. In fact, health care providers screened out potential cases if they believed the FCG to be experiencing great distress. Fourth, health care providers often indicated that they did not have time to help recruit, as their workloads were too demanding. Fifth, health care providers did not always know who was providing care in the home, given that it is the family member, not the FCG, who is the client in the health care encounter. This occurs despite the fact that both the family member and the family are philosophically understood to be the unit of care in palliative situations. In some cases, this necessitated having to first get consent from the family member to identify and approach FCGs about the research. The idea that FCGs are often overlooked is reinforced by them not being viewed as legitimate recipients of care, suggesting that FCGs might be more 'at risk' or vulnerable than we understand.

In addition to these systemic barriers, recruitment is affected by participant-specific factors. For example, given that caregiving work in the home is invisible, many FCGs do not identify as a caregiver per se, nor do they necessarily feel entitled to receive support (Stajduhar, 2003). This is problematized by the fact that services are provided on the basis of family member need only, given that FCGs are not formally recognized as clients in the health care situation. Further, recruiting participants is difficult because some FCGs are uncomfortable with the research topic or process. As suggested, not all FCGs and family members are accepting of a terminal prognosis, nor are they comfortable speaking about dying, death or the stresses of caregiving. Finally, recruiting is difficult because FCGs lack time and energy. These latter issues are associated with poor retention of FCGs in longitudinal research, wherein they withdraw from the research if caregiving becomes too burdensome or stressful (Sherman, et al., 2005). Fortunately this has not been my experience (Donovan, et al., 2010; Donovan and Williams, 2011; Donovan and Williams, 2015).

Managing access and recruiting issues

A number of strategies were implemented to address the difficulties with recruitment. The most important step was trust-building with both health care providers and FCGs. Specific to health care providers, seeking their input to best determine how they can be served by the research allowed them to recognize the integrity of the project and see how the work is meant to inform and, ultimately, assist them in caring for the family member, the FCG and the larger family unit. For example, health care providers were invited to collaborate on the research through the identification of specific questions they wanted answered. Training and recruitment information packages, including recruitment cards and brochures containing details about the research and the researcher's contact information, were provided to limit the demands on their time. Further, offers were regularly made to share the results of the research program with their care teams via presentations and lay reports. Finally, in return for participant recruitment, offers were made to provide monetary donations toward in-service training specific to palliative care. One of the harder realizations was to accept that not all health care organizations are supportive of research; while time constraint was a significant issue, many are not research-oriented.

With FCGs, trust-building entailed taking the time to explain the research and allowing time for them to ask questions. Further, trust was reinforced by being sincere about the promises made (e.g., reimbursement, participant-driven strategy, etc.), being a good listener, and knowing if, when, and how to appropriately probe responses. To illustrate, several steps were involved in the recruitment process: the research was first introduced to the FCG via a recruitment postcard delivered by visiting formal providers, in which potential participants, if interested, were asked to call the researchers for further information. If FCGs agreed, researchers would then meet with them in person to go over the LOI/CFs, allowing them time to ask questions, hear sample questions, and clarify all aspects of the process. Only then did I make arrangements for the interview – either offering to return at their convenience or proceeding with the interview if the FCG preferred. Following the interview, researchers would reconfirm that they were comfortable with what had occurred. In this way, researchers were certain that FCGs made an informed decision about participating in the research. This approach was critical in developing trust in the relationship and in avoiding potential misunderstandings. It was particularly appropriate to use when other people, such as health care providers and research assistants, were involved with recruitment, and especially so in cross-cultural contexts, given the risk of misunderstandings or miscommunication. In cross-cultural contexts a cultural entrepreneur was always employed to minimize these risks. In addition to building trust, this investment of time and energy early in the researcher-participant relationship contributed to participant retention, insofar as FCGs understood the procedures and the participant-driven nature of the process. These measures not only demonstrate professionalism, but are essential to the ethical conduct of qualitative research in general.

To widen the potential sample and further facilitate recruitment, it was also necessary to be flexible in the research design, as was earlier discussed in connection with the 'participant-driven' data collection strategy. As an example of this flexibility, it was determined that it was often preferable for FCGs, family members and health care providers alike to refer to the nature of the illness as 'serious or life-threatening,' instead of 'palliative.' The REB had no concerns specific to the changes of the wording used in recruitment and all other relevant research materials, from 'palliative' to 'serious or life-threatening', given the context of the researchers' commitment to transparency and fully informed consent. To increase participant numbers, FCG participants were also accessed via acute or chronic care caseloads, as opposed to the palliative caseloads only. Finally, FCGs appreciated gestures of gratitude, through monetary honorariums and other small tokens, such as pots of soup, cards, flowers, and perennial shrubs, to compensate them for their contribution to the research. These various recruitment strategies were implemented with persistence, which was the single underlying characteristics of all recruitment efforts.

The preceding two themes demonstrate the ethical and logistical challenges involved in palliative care research, many of which are also emotional in nature. However, I have mainly focused on the perspectives of FCGs (and, to some degree, family members) as participants, REBs and health care providers as institutionally based facilitators of the research. Yet, this chapter will also reveal that qualitative researchers are emotionally affected by the work they do, addressed in the final theme below.

Emotion work in palliative care research

The nature and process of this research can be uncomfortable and upsetting for both FCGs and health care providers. However, as researchers we are also affected, as we are not detached from the work we do. This is true of any qualitative research, given the intimate interaction between the researcher and participants; however, discussions concerning the emotional impact of the process on researchers are limited (Bondi, 2005). The concept of emotion work represents specific concern for and attention to the affective elements of everyday life and this intricate relationship to one's socio-spatial relations (Bondi, et al., 2005). In this section, the emotion work is discussed in relation to research with palliative FCGs.

There are two areas of intense emotional response: the first concerns the participant-researcher relationship, and the second, the impact of the research on the researcher. The ways in which participants are treated is central to this relationship, as has already been noted. In addition to being ethical, there is a need to be professional and authentic. This means being sensitive to FCGs needs and avoiding becoming a burden. One strategy that was used to maximize the potential of this relationship was that the researchers conducted the interviews themselves, being sure to be consistently matched to the same FCG. That is, a single FCG saw the same researcher for all encounters, whether making appointments over the phone or carrying out an interview in person.

There are many ways in which the researcher is impacted by the research, as both a qualitative researcher and palliative care researcher specifically. One is the researcher's emotional reaction to the research. For example, the palliative situation immediately conjures up feelings of sympathy; it is difficult not to feel for the FCG. As well, having previously experienced caregiving and bereavement, researchers are reminded of our own feelings – both positive and negative – when hearing FCGs talk about their situations. Further, guilt was felt for asking FCGs to participate in the research when their time, energy and emotions were already stretched. Consequently, the research was approached with caution and some uncertainty, in terms of how FCGs would react to the research and the ways in which they could be affected by participation. Learnings suggest the need to set these feelings aside when approaching the research enterprise.

Given that FCGs were asked to share their experiences, in terms of the scope of care and the meaning it had for them, interviews were highly emotional. Bearing witness to their feelings as they journeyed through these experiences (literally, given the longitudinal approach, and figuratively, in hearing them retell their experiences), was both powerful and rewarding. However, it was also very troubling to recognize that, for some participants, the researchers were the only people with whom these feelings were shared. This relationship has the potential to create situations which may reinforce the power differentials between the participants and the researchers, through feelings of dependency and overreliance. This speaks to the importance of maintaining boundaries in the research relationship, as we cannot confuse our professional role as researchers with that of a professional counselor or simply a friend, even if it appears to be what is needed by the FCG. In the following section the strategies found most useful in managing the challenges of emotion work are discussed.

Managing the challenges of emotion work

A number of strategies were used to address these emotional issues, some of which have already been alluded to in the previous themes. Specific to FCGs, it was critical to have a good understanding of caregiving; dying, death and bereavement;

depression; and grief to anticipate and appropriately approach sensitive topics, as well as to recognize and respond to distress in the interviews. Furthermore, it was sometimes necessary and helpful to share with FCGs personal experiences or knowledge of the extant literatures specific to caregiving and bereavement, simply to assure them that what they were feeling was not unique or inappropriate. As well, in each interaction, researchers came prepared with a list of resources in the community that could be accessed by FCGs to help deal with the issues they were facing. Many times, FCGs were unaware that these services existed or simply did not know where or how to access them.

As has also been alluded to, it has been my and my research team's experience that FCGs found the process of participating in the research very therapeutic. First, the research process helped affirm the altruism of these caregivers. Further, the FCGs valued the opportunity to contribute to supporting FCGs in the future by sharing their own experiences. In fact, similar to other research findings (Gysels, et al., 2008b), most, if not all of the FCGs who participated in the research program have indicated helping others is the primary reason for participating given that they wished they had more support themselves. FCGs also indicated that they truly valued the opportunity to share their experiences with an attentive and non-judgmental listener. The privacy they were accorded in one-to-one interviews enabled them to speak freely, emotionally if necessary, and put matters into perspective. This was enhanced by the longitudinal component of the research, often which captured the range of emotions in relation to caregiving over time. However, particular attention was given to both verbal and non-verbal cues that could indicate FCG distress and the possibility that a particular question or the interview in general was too upsetting. Certain FCGs clearly wanted the research relationship to develop into a long-term friendship; this was managed by continuously bringing these FCGs back to the fact that I was engaged in a research enterprise, reminding them that the project would be eventually coming to an end.

Several strategies were used to self-manage the emotional impact of this research. For example, it was helpful to have measures in place to brainstorm and debrief with senior colleagues and fellow researchers, in confidence, about the challenges and feelings that were experienced, specific to the research or from the caregiving situations directly. Another strategy was to schedule some alone or one-on-one time after the interviews to allow time to process the information that was collected. Thoughts and feelings were released through writing in field notes, journals and diaries, all of which are expected venues for rigorous qualitative research, as well as through physical activity, and sometimes crying. Using these strategies in combination was particularly effective for self-managing emotions, particularly when the FCG respondents were so forthcoming with their experiences.

It was also necessary to be persistent in terms of establishing relationships and in recruiting participants. Persistence was also required in addressing emotions that were revealed throughout the process. Sorting out feelings was necessary to remain grounded, albeit sensitive, to research participants; this was also necessary for determining how personal emotions might be impacting the research process.

Finally, to ensure that the research outcomes produced by the research process were widely distributed, a comprehensive knowledge translation (KT) strategy was developed and is still being undertaken. These research outcomes are seen as the ultimate end to the research program's work and make the research investment worthwhile from both a professional and emotional perspective. This KT strategy creates awareness about the issue of palliative family caregiving and the contribution of social scientists within the realm of health care practice and policy specifically. This includes producing lay reports for participants and relevant health care providers, as well as presenting and publishing via traditional academic outlets. This approach will not only increase the likelihood that research findings can be applied directly to service improvement, but illustrates the necessity and importance of an interdisciplinary perspective in health services research. One highlight has been multiple invitations to speak to parliament in Ottawa, regarding compensation for family palliative caregivers.

Discussion and conclusion

This chapter has demonstrated how palliative care research is infused with ethical, practical, and emotional challenges that are interconnected in complex and substantial ways. By sharing experiences specific to conducting palliative care research, this chapter illustrates how the challenges of conducting sensitive qualitative research can be managed to produce meaningful health-related research. This may be useful to both new and experienced qualitative researchers, including both social scientists and others who are engaged in similarly sensitive qualitative and palliative care research.

The ethical issues that feature prominently as challenges in qualitative palliative care research are addressed by limited literature. There is currently a disconnect between REBs and FCGs with respect to perceptions of vulnerability and distress (Buckle, et al., 2010). In particular, issues of risk are overemphasized without adequate consideration of the benefits of participation and the contribution to well-being and adjustment in bereavement (Buckle, et al., 2010; Casarett and Karlawish, 2000; Emanuel, et al., 2004; Gysels, et al., 2008b). I agree that distress is not caused by the research per se, as FCGs are sharing their actual lived experiences; it is the sharing that must be managed by the researcher (Buckle, et al., 2010). The interview provides a venue through which to freely, honestly and fully express distress and grief; this results in an unintended, but decidedly therapeutic, benefit for FCGs. As suggested, strategies that privilege complete and ongoing consent and 'participant-driven' data collection can empower FCGs to make informed and voluntary decisions about participation in research, whilst contributing to therapeutic outcomes in the process.

The emphasis on issues pertaining to vulnerability and protection of FCGs are paradoxical, given the extent to which FCGs themselves assume a wide range of caregiving and decision-making responsibilities with little choice or support (Stajduhar, 2003). Buckle and colleagues (2010) provide perspective by pointing out that in most western societies, bereaved family members are believed to be strong enough to make sound decisions regarding funeral arrangements and estate matters, and return to work within a few days of the death of an immediate family member. Yet when it comes to research, FCGs are somehow regarded as being too vulnerable and incapable of making autonomous and informed decisions about involvement. Overprotection based on a priori assumptions about vulnerability by well-intentioned gatekeepers of ethical and service-based institutions can serve to deny FCGs a much-deserved and necessary voice in palliative care research; to do so is both oppressive and unethical. The failure to consider them as autonomous beings capable of making informed decisions further marginalizes them and, I argue, perpetuates the shroud of invisibility under which caregiving is already physically and emotionally experienced. This suggests that FCGs might be more in need of protection from the systems which render them invisible than they do from the researchers who are trying to help them.

I am not suggesting that researchers do not need to be held to a high ethical standard – quite the contrary, as indicated in the suggestions to manage the noted research challenges. It is possible, however, that this research topic still holds discomfort for many, given our death-denying culture (Northcott and Wilson, 2008), resulting in reluctance to approve research or to trust what is being proposed by researchers. I suspect that this angst is experienced by researchers, participants, and REBs with similarly sensitive qualitative research and work with difficult or vulnerable populations. To break down these barriers and facilitate the production of high-quality research, it is imperative for qualitative researchers to take more time to share their experiences concerning challenges and the emotional impact of their work. This will not only inform the shape of future research, but will provide REBs and potential collaborators with greater knowledge about the benefits of the research so that they might be more supportive of the pursuit for evidence which is necessary to inform health services and health policy.

References

Addington-Hall, J. M., 2007. Introduction. In Addington-Hall, J. M., Bruera, E., Higginson, I., and Payne, S. (eds.) *Research methods in palliative care*. New York: Oxford University Press. pp. 1–9.

Agrawal, M., 2003. Voluntariness in clinical research at the end of life. *Journal of Pain and Symptom Management*, 25(4), pp. S25–S32. Available at: http://www.jpsmjournal.com/article/S0885-3924%2803%2900057-5/abstract.

Baxter, J., and Eyles, J., 1997. Evaluating qualitative research in social geography: establishing 'rigour' in interview analysis. *Transactions of the Institute of British Geographers*, 2(4), pp. 505–525. doi:10.1111/j.0020-2754.1997.00505.x.

Bondi, L., 2005. The place of emotions in research: From partitioning emotion and reason to the emotional dynamics of research relationships. In Davidson, J., Bondi, L., and Smith, M. (eds.) *Emotional geographies*. Hampshire, England: Ashgate Publishing. pp. 231–246.

Bondi, L., Davidson, J., and Smith, M., 2005. Introduction: Geography's 'emotional turn'. In Davidson, J., Bondi, L., and Smith, M. (eds.) *Emotional geographies*. Hampshire, England: Ashgate Publishing. pp. 1–16.

Brown, M., 2003. Hospice and the spatial paradoxes of terminal care. *Environment and Planning, A*, 35(5), pp. 833–851. doi:10.1068/a35121.

Buckle, J. L., Dwyer, S. C., and Jackson, M., 2010. Qualitative bereavement research: incongruity between the perspectives of participants and research ethics boards. *International Journal of Social Research Methodology*, 13(2), pp. 111–125. doi:10.1080/13645570902767918.

Carstairs, S., 2010. Raising the bar: A roadmap for the future of palliative care in Canada. The Senate of Canada (Federal Government of Canada). Available at: http://www.chpca.net/media/7859/Raising_the_Bar_June_2010.pdf.

Casarett, D.J., and Karlawish, H. T., 2000. Are special ethical guidelines needed for palliative care research? *Journal of Pain and Symptom Management*, 20(2), pp. 130–139. Available at: http://www.jpsmjournal.com/article/S0885-3924%2800%2900164-0/abstract.

Crossley, M., 2007. Evaluating Qualitative Research. In Addington-Hall, J. M., Bruera, E., Higginson, I., and Payne, S. (eds.) *Research Methods in Palliative Care*. New York: Oxford University Press. pp. 181–190.

Cutcliffe, J. R., 2003. Reconsidering reflexivity: introducing the case for intellectual entrepreneurship. *Qualitative Health Research*, 13(1), pp. 136–148. doi:10.1177/1049732302239416.

Donovan, R., and Williams, A., 2011. Shifting the burden: The effects of home-based palliative care on family caregivers in rural areas. In Kulig, J., and Williams, A.M. (eds.) *Health in Rural Canada*. Vancouver, British Columbia: University of British Columbia Press. pp. 316–333.

Donovan, R., and Williams, A., 2013. Caregiving as a Vietnamese Tradition: 'It's like eating, you just do it'. *Health and Social Care in the Community*, 23(1), pp. 79–87.

Donovan, R., and Williams, A., 2014. Caregiving as a Vietnamese Tradition: 'It's like eating, you just do it'. *Health and Social Care in the Community*, 23(1), pp. 79–87. doi:10.1111/hsc.12126

Donovan, R., Williams, A., Stadjuhar, K., Brazil, K., and Marshall, D., 2010. The influence of culture on home-based family caregiving at end-of-life: A case study of Dutch reformed family caregivers in Ontario, Canada. *Social Science and Medicine*, 72, pp. 338–346. doi:10.1016/j.socscimed.2010.10.01.

Dowling, R., 2005. Power, subjectivity, and ethics in qualitative research. In Hay, I. (ed.) *Qualitative research methods in human geography*. 2nd ed. Australia: Oxford University Press. pp. 19–29.

Duke, S., and Bennett, H., 2010. A narrative review of the published ethical debates in palliative care research and an assessment of their adequacy to inform research governance. *Palliative Medicine*, 24(2), pp. 111–126. doi:10.1177/0269216309352714.

Dyck, I., 1999. Using qualitative methods in medical geography: deconstructive moments in a subdiscipline? *Professional Geographer*, 51(2), pp. 243–253. doi:10.1111/0033-0124.00161.

Dyck, I., 2001. Centring the home in research. *Canadian Journal of Nursing Research*, 33(2), 83–86. Available at: http://digital.library.mcgill.ca/cjnr/search/issue.php?issue=CJNR_Vol_33_Issue_02.

Dyck, I., Kontos, P., Angus, J., and McKeever, P., 2005. The home as a site for long-term care: Meanings and management of bodies and spaces. *Health and Place*, 11, pp. 173–185. doi:10.1016/j.healthplace.2004.06.001.

Emanuel, E. J., Fairclough, D. L., Wolfe, P., and Emanuel, L. L., 2004. Talking with terminally ill patients and their caregivers about death, dying, and bereavement: Is it stressful? Is it helpful? *Archives of Internal Medicine*, 164, pp. 1999–2004. doi:10.1001/archinte.164.18.1999.

Grande, G., Stajduhar, K., Aoun, S., Toye, C., Funk, L., Addington-Hall, J., Payne, S., and Todd, S., 2009. Supporting lay carers in end of life care: current gaps and future priorities. *Palliative Medicine*, 23, pp. 339–344. doi:10.1177/0269216309104875.

Gysels, M., Shipman, C., and Higginson, I. J., 2008a. Is the qualitative research interview an acceptable medium for research with palliative care patients and carers? *BMC Medical Ethics*, 9, 7. Available at: http://www.biomedcentral.com/1472-6939/9/7.

Gysels, M., Shipman, C., and Higginson, I.J., 2008b. 'I will do it if will help others': Motivations among patients taking part in qualitative studies in palliative care. *Journal of Pain and Symptom Management*, 35(4), pp. 347–355. Available at: http://www.jstor.org/stable/27720367.

Johnson, B., and Clarke. J., 2003. Collecting sensitive data: the impact on researchers. *Qualitative Health Research*, 13(3), pp. 421–434. doi:10.1177/1049732302250340.

Kearns, R., and Moon, G., 2002. From medical to health geography: novelty, place and theory after a decade of change. *Progress in Human Geography*, 26(5), pp. 605–625. doi:10.1191/0309132502ph389oa.

Koenig, B. A., Back, A. L., and Crawley, L. M., 2003. Qualitative methods in end-of-life research: recommendations to enhance the protection of human subjects. *Journal of Pain and Symptom Management*, 25(4), pp. S43–S52. Available at: http://www.jpsm-journal.com/article/S0885-3924%2803%2900060-5/abstract.

Miller, K., McKeever, P., and Coyte, P.C., 2003. Recruitment issues in healthcare research: the situation in home care. *Health and Social Care in the Community*, 11(2), pp. 111–123. doi:10.1046/j.1365-2524.2003.00411.x.

Milligan, C., 2003. Location or dis-location? Towards a conceptualization of people and place in the care-giving experience. *Social and Cultural Geography*, 4(4), pp. 455–470. doi:10.1080/1464936032000137902.

Milligan, C., Atkinson, S., Skinner, M., and Wiles, J., 2007. Geographies of care: a commentary. *New Zealand Geographer*, 63(2), pp. 135–140. doi:10.1111/j.1745-7939.2007.00101.x.

Northcott, H. C. and Wilson, D. M., 2008. *Dying and death in Canada*. 2nd ed. Peterborough, Canada: Broadview Press.

Payne, S. A., and Turner, J. M., 2008. Research methodologies in palliative care: a bibliometric analysis. *Palliative Medicine*, 22, pp. 336–342. doi:10.1177/0269216308090072.

Romanow, R. J., 2002. *Building on values: the future of health care in Canada*. Ottawa: Commission on the Future of Health Care in Canada. Available at: http://publications.gc.ca/collections/Collection/CP32-85-2002E.pdf.

Sherman, D. W., Beyers McSherry, C., Parkas, V., Ye, X. Y., Calabrese, M., and Gatto, M., 2005. Recruitment and retention in a longitudinal palliative care study. *Applied Nursing Research*, 18, pp. 167–177. Available at: http://www.appliednursingresearch.org/article/S08971897%2805%2900037-6/abstract.

Stajduhar, K. I., 2003. Examining the perspectives of family members involved in the delivery of palliative care at home. *Journal of Palliative Care*, 19(1), pp. 27–35. Available at: http://www.ncbi.nlm.nih.gov/pubmed/12710112.

Stajduhar, K. E., and Cohen, R., 2009. Family caregiving in the home. In Hudson, P., and Payne, S., eds. *Family carers in palliative care*. New York: Oxford University Press. pp. 149–168.

Steinhauser, K. E., Clipp, E. C., Hays, J. C., Olsen, M., Arnold, R., Christakis, N. A., Hoff Lindquist, J. and Tulsky, J. A., 2006. Identifying, recruiting, and retaining seriously ill patients and their caregivers in longitudinal research. *Palliative Medicine*, 20, pp. 745–754. doi:10.1177/0269216306073112.

Wiles, J., 2003. Daily geographies of caregivers: mobility, routine, scale. *Social Science and Medicine*, 57, pp. 1307–1325. doi:10.1016/S0277-9536(02)00508-7.

Williams, A., 2002. Changing geographies of care: Employing the concept of therapeutic landscapes as a framework in examining home space. *Social Science and Medicine*, 55, pp. 141–154. Available at: http://www.ncbi.nlm.nih.gov/pubmed/12137183.

Williams, A., 2004. Shaping the practice of home care: Critical case studies of the significance of the meaning of home. *International Journal of Palliative Nursing*, 10(7), pp. 333–342. Available at: http://www.internurse.com/cgibin/go.pl/library/article.cgi?uid=14575;article =IJPN_10_7_333_342;format=pdf.

Williams, A., and Donovan, R., 2007. Making Home Space Therapeutic for Palliative Patients and Their Caregivers. In A. M. Williams, ed. *Therapeutic Landscapes: Advances and Applications*. Burlington, VT: Ashgate Publishing. pp. 199–218.

Williams, B. R., Woodby, L. L., Bailey, F. A. and Burgio, K. L., 2008. Identifying and responding to ethical and methodological issues in after-death interviews with next-of-kin. *Death Studies*, 32(3), pp. 197–236. doi:10.1080/07481180701881297.

World Health Organization, 2007. WHO definition of palliative care. Available at: http://www.who.int/cancer/palliative/definition/en/.

14 Informal caregiving on the move

Examining the experiences of Canadian medical tourists' caregiver-companions from patients' perspectives

Valorie A. Crooks, Victoria Casey and Rebecca Whitmore

Reflecting on praxis

This chapter speaks to the value of conducting secondary analyses in qualitative research in general, and qualitative health geography specifically. In it we present the findings of a study exploring the decision-making processes of Canadian medical tourists. However, although that is what the study probed, the analysis presented here provides a first glimpse into the experiences of the friends and family members who accompany them abroad (i.e., their caregiver-companions). In other words, while we set out to understand Canadian medical tourists' decision-making processes, we ended up unexpectedly learning about this informal caregiver group. In fact, we were struck by the potential for this secondary analysis during data collection when, in the very first interview we conducted, a former medical tourist talked about how his wife had accompanied him abroad and while in India she became ill and required emergency medical attention. Hearing this experience made us wonder what roles and responsibilities this group was taking on, if they may be experiencing caregiver burden given the unique and demanding transnational care context, and if the data we were gathering from medical tourists could offer any meaningful insights. What we present in this chapter shows that we learned much about this important yet mostly invisible stakeholder group. If we were not open to pursuing new avenues of investigation in the current data set via secondary analysis, then we may never have become attuned to the vital roles and responsibilities taken on by the friends and family members who accompany medical tourists abroad.

As we stated above, from the first interview we conducted with a former Canadian medical tourist in the study reported on herein we began to develop an interest in caregiver-companions. This interest grew throughout the course of the study as we learned more about this stakeholder group, leading us to pursue this secondary analysis, and now we are conducting a new large study specifically about informal caregiving in medical tourism. The purpose of our new study is to gain an understanding of the experiences of Canadian medical tourists' informal caregiver-companions through gathering and analyzing their first-hand accounts and those of other key stakeholders. Unlike the secondary analysis reported on in this chapter, this new study is enabling us to collect new insights specific to this stakeholder group so that we can pursue primary analyses central to the study objectives. Our specific objectives are to (1) determine how caregiver-companions understand and respond to the health and safety risks associated with medical tourism, both for themselves and for the medical tourist; (2) explore how caregiver-companions characterize their roles in relation to the medical tourist; and (3) examine the ethical and practical dimensions of the responsibilities they assume for the care of the medical tourist and themselves. This new study is built directly from the insights gleaned from the secondary analysis presented in this chapter and has been funded by an Operating Grant from the Canadian Institutes of Health Research. We recently published our first papers from this study (Casey, et al., 2013a; 2013b), both of which confirm various aspects of the findings that emerged from the secondary analysis presented here.

The secondary analysis presented in this chapter provides a foundation for the ongoing research described above. If participants in the original study had not mentioned the friends and family members who accompanied them abroad, we likely would not have pursued this line of inquiry. This secondary analysis also highlights how novel and unexpected themes emerge from qualitative data sets that provide valuable insight into phenomena, even though they do not address the stated purpose of the original study. Qualitative researchers regularly observe such themes arising in their interview data. Heaton (2008) termed this type of secondary analysis as 'supplementary analysis,' wherein researchers complete a more detailed analysis of an issue emerging from the original data set that was not central to the study objectives. Problematic limitations can arise in secondary analysis when data are analyzed by researchers who lack knowledge of how data were generated, and when the data may not be suited for secondary analysis (as in the case of ethnographies where researchers engaged in immersive fieldwork) (Heaton, 2008; Irwin, 2013). Fortunately in the case of our own analysis, the same lead investigator that gathered the data was involved in the secondary analysis. However, there are significant benefits to secondary analysis for all qualitative researchers, including qualitative health geographers, such as the 'fresh eyes' new researchers may bring to the data as a result of their lack of involvement in data creation (Irwin, 2013). We benefitted from new perspectives in this analysis since two of us were not members of the team involved in gathering the original data set.

Introduction

Medical tourism involves patients intentionally leaving their country of residence outside of established cross-border care arrangements to access non-emergency medical interventions, often surgeries, which they typically pay for out-of-pocket (Ehrbeck, et al., 2008). Injured vacationers seeking care, people accessing health tourism services such as spa treatments while abroad, and expatriates receiving care in the region in which they live are not medical tourists by this definition (Bertinato, et al., 2005; Pachanee and Wibulpolprasert, 2006). Some medical tourists spend part of their postsurgical recovery period abroad in resorts or hotels that partner with hospitals treating international patients (Garcia-Altes,

2005; Whittaker, 2008). Established medical tourism destination nations include India, Singapore, Cuba, Costa Rica, Germany, and Poland, while a number of other countries are emerging as medical tourism destinations, especially throughout Central America and the Caribbean. This chapter reports on a study that involved interviews with Canadian medical tourists and addresses the support they described receiving from friends and family members while at home and abroad. Semi-structured interviews offered a highly suitable method to explore the decision-making experiences of individuals engaging in this emerging health phenomenon, and secondary analysis allowed us to focus on a previously unexamined aspect of medical tourism: the practice of informal caregiving.

Much existing research on medical tourism is focused on the system level. It raises a range of concerns that hold negative implications: (1) use of public health system resources for medical tourism in destination countries can create health inequities, particularly in low and middle income countries (Mudur, 2004; Sen Gupta, 2008); (2) poor malpractice protections in destination nations can leave international patients without legal recourse (Cortez, 2008); (3) the lure of practicing in high-end medical tourism facilities may exacerbate internal brain drain in destination countries (Connell, 2008); and (4) medical tourism may create or exacerbate two-tier care in departure (i.e., patients' home nations) and destination countries (Garud, 2005; Whittaker, 2008). At the same time, medical tourism has been heralded as a solution to global health problems by (1) providing foreign investment to destination nations' health systems (Garcia-Altes, 2005; Connell, 2008); (2) lessening the international migration of health workers from the health systems of low and middle income countries by providing new employment opportunities (Lautier, 2008; Whittaker, 2008); (3) offering new forms of medical care to citizens of destination countries and often affordable care for international patients (York, 2008); and (4) serving as a way to lessen wait list bottlenecks in departure countries (Botten, et al., 2004; Leahy, 2008). While it is important to study these and other system-level impacts of medical tourism, we must also gain a better understanding of how this practice is experienced by those involved in order to develop appropriate health system and health policy responses to medical tourism (Crooks, et al., 2010). At present, we do not have an adequate understanding of the lived dimensions of medical tourism, including how it is experienced by the full range of stakeholders directly and indirectly involved in and impacted by this practice.

Canadians are among those international patients participating in medical tourism by purchasing private health care abroad. Although there are no reliable quantitative data on outbound medical tourism from Canada, it is safe to say that, at minimum, several thousand Canadians are going abroad each year as medical tourists for plastic, transplant, orthopedic, dental, experimental, and other procedures (Yelaja, 2006; Laidlaw, 2008), and some media reports put this number in the tens of thousands. In recent years, key developments such as plans to make Canada a medical tourism destination (Fowlie, 2010; Weber, 2010), rallies to have provincial ministries cover the costs of surgeries undertaken abroad by medical tourists (CBC, 2010) and the spread of the NDM-1 superbug to Canada by medical tourists (Westhead, 2010; White, 2010) have demonstrated that medical tourism is a highly contested and risky practice. These developments have also drawn attention to a wide range of evidence and policy gaps regarding Canadians' participation in medical tourism and the implications this practice holds for the Canadian health care system.

Alongside practices such as health worker migration, physician voluntourism, cross-border care and medical outsourcing, medical tourism is an example of a highly spatial globalized health service that involves the transnational movement of people, providers, and health information (Wibulpolprasert, et al., 2004; Hopkins, et al., 2010). The international literature consistently refers to a number of medical tourism stakeholders: international patients, health care providers and administrators in destination countries, health care providers and administrators in patients' home countries, brokers and facilitators, and insurance agents. These groups are cited as having primary involvement in delivering and receiving medical tourism services, and valuable research focused on their interests and experiences is beginning to emerge (Crooks, et al., 2010). In this chapter we examine the experiences of a far less considered medical tourism stakeholder group: medical tourists' informal (i.e., unpaid) caregiver-companions (e.g., friends, family members). We focus here specifically on Canadian medical tourists' caregiver-companions, using secondary analysis as a useful method to gain insight into this unstudied group.

In this chapter we show that caregiver-companions typically stay in the hospital room and recovery hotel or resort with medical tourists, and provide some vital forms of informal care, such as hands-on care, symptom management, coordination and emotional support. Given that caregiver-companions are providing care that is seemingly integral to the success of medical tourism suggests that they are, in effect, acting as shadow workers (i.e., unpaid, untrained and unrecognized care providers) in the delivery of this global health service (Armstrong, et al., 2001). Meanwhile, we know almost nothing about who they are and the types of ethical and practical responsibilities they take on or the implications that providing such care holds for their own health. Secondary analysis of semi-structured interviews with Canadian medical tourists both alerted us to the existence of this group and provided insight into the roles they play. Health geographers interested in qualitatively studying groups like caregiver-companions, whose existence is not well-known or are difficult to locate, may find it valuable to consider interviews with other, more accessible stakeholder groups as a way to gather information about the target group.

Study overview

The insights we share in this chapter were gleaned from a secondary analysis of an exploratory study investigating the decision-making processes and experiences of Canadian medical tourists. The original study involved interviews with Canadian medical tourism facilitators (i.e., agents who make bookings for prospective medical tourists) and former Canadian medical tourists. Here we draw on the interviews conducted with former Canadian medical tourists to shed light on the experiences of the caregiver-companions who accompanied them abroad. We have published the findings of these interviews quite extensively (e.g., Crooks, et al., 2012; Johnston, et al., 2012; Snyder, et al., 2012), so here we provide just a brief overview of the study and secondary analysis as further details can be sought out in our other publications. This is a secondary analysis in that the themes we examine regarding caregiver-companions emerged from the data but do not in any way address the study's objectives and purpose.

In 2010 we conducted 32 semi-structured phone interviews with Canadians who had previously gone abroad as medical tourists for surgical interventions. We identified these individuals using multiple strategies, including advertising in print media and posting information about our study on medical tourism online forums. The phone interviews typically lasted 1 to 1.5 hours and were conducted after participants signed a consent form. During the interviews we probed participants' motivations for going abroad, information seeking processes, assessment of risks and receipt of postoperative care upon return home.

The interviews were transcribed verbatim. An initial coding scheme was created following transcript review by the lead investigators, where it was agreed that enough information was shared about the participants' caregiver-companions that this should be coded for. Coding was undertaken using the NVivo data management program. Following coding, data pertaining to caregiver-companions were extracted and reviewed by the authors in order to identify themes and determine whether or not a secondary analysis of these data was warranted. We each read these data extracts and agreed upon three themes and their scope through an iterative process of data review and discussion. We expand upon the three themes identified in our secondary analysis in the section that follows.

Secondary analysis findings

Twenty-one of the 32 former Canadian medical tourists we interviewed had travelled abroad with caregiver-companions. Those 21 collectively took 22 medical tourism trips: nine participants went to India, five went to the United States, three went to Cuba, and the remaining participants travelled to Germany, Mexico, South Africa and Poland. Nearly two thirds of participants (n = 13) were accompanied by a married partner, while the others were accompanied by an unmarried partner (n = 3), a friend or friends (n = 2), a sibling (n = 2), or other family members (n = 2). After reflecting on their experiences, the majority (n = 16) reported that they would recommend that medical tourists travel with a caregiver-companion, 11 said that it could be useful to do in some contexts (e.g., if it is affordable or if the patient is not able to pay to recover at a clinic that provides all the necessary care), three advised against it, and three were indifferent.

Using thematic analysis we identified three main themes that best characterize the breadth of information the medical tourists shared about their caregiver-companions: (1) caregiver-companions have particular types of caregiving responsibilities, (2) care is provided in a variety of spaces, and (3) caregiver-companions negotiate relationships in caregiving while abroad. We explore these themes in this section, providing quotes throughout in order to let the participants' voices speak directly to the issues.

Caregiving responsibilities

Caregiver-companions have a variety of caregiving responsibilities according to medical tourists. The reported responsibilities can be divided into three main types: emotional care, hands-on care, and care logistics. The emotional care delivered by caregiver-companions can come in a variety of forms. Speaking of her experiences of travelling accompanied, a participant described the emotional support provided by her caregiver-companion: 'I'm quite independent but I would say emotionally you need to have somebody you can count on to be with you because [the country is] different. . . . It's not your home. It's not your . . . language . . . somebody that is just a little bit insecure can get a *lot* insecure.' One form of emotional support is the provision of moral support, whereby caregiver-companions can help medical tourists make the decision to have the procedure in the first place and later reassure them of the soundness of that decision, maintain an extended support network through communication with family and friends back home while abroad, and act as a familiar element in an unfamiliar environment, whereby they help the medical tourist negotiate the foreignness of the country and the clinic. Emotional care can also be passive when caregiver-companions act as someone to whom the medical tourist can complain, whose presence offers them company and a feeling of safety, and who can empathize with their experiences by 'going through the whole thing with [the patient]'. Caregiver-companions can further provide emotional care by inspiring the feeling of being cared for through purchasing small gifts such as flowers, movies or even a simple pizza slice. Emotional care is very commonly provided by caregiver-companions in one or many forms at home and abroad.

Hands-on care, although a primary responsibility of health care practitioners in destination facilities, is taken on by some caregiver-companions. This can include clinical tasks such as monitoring the patient's liquid intake, reminding the patient to take medication or administering medications, reminding patients and helping them to follow postoperative orders, and taking charge as complications occur. One participant listed the tasks of this nature that were performed by her caregiver-companion: 'He administered my pain medication. He monitored my, liquid intake, which is very, very important post . . . surgery. . . . I actually fell in the bathroom so he . . . bandaged up my entry wounds from the surgery.' Some forms of hands-on care are done for the comfort of patients, such as finding food that will ease discomfort or accompanying them to meetings with a surgeon. Participants who travelled with caregiver-companions frequently reported that their companions helped them with everyday tasks such as getting dressed, getting in and out of bed, getting to the bathroom, carrying items from across the hospital room and finding things from outside of the hospital, such as food. 'If I needed anything, my husband, he's the gopher; he'd always go for it for me,' one participant reported. Caregiver-companions also adopt tasks that promote patient safety during the travel back home. If driving, they can stop the car frequently to allow for rest breaks, and if flying, they can make sure the patient wears circulation socks. Both of these measures can help to prevent deep vein thrombosis during the recovery period. They also typically carry patients' luggage. Upon return home, the caregiver-companion can continue hands-on tasks such as helping them with the household chores, bringing them food and keeping them comfortable. Overall, hands-on care is a practical responsibility that emphasizes the patient's healing and physical comfort.

A third caregiving responsibility taken on by caregiver-companions pertains to arranging the logistics of care. This primarily involves ensuring that the medical tourism journey goes smoothly for the patient. One participant described the logistical coordination undertaken by his caregiver-companion like this: 'She [kept] all the files, the information, to go here, go there. . . . [I] just didn't have the wherewithal to . . . navigate and negotiate and so she did all of that stuff.' Such logistical coordination can encompass the management of patient transportation, which can include driving the patient to the airport in either the destination country or country of origin (or driving them directly to and from the destination), making sure that the airplane transfers are successful, arranging for a taxi to and from the hospital or driving within the destination country. Planning, including creating an itinerary and ensuring the patient follows the schedule, is also part of the logistical coordination responsibilities taken on by caregiver-companions. Furthermore, they often handle the logistics of storing and transferring paperwork and may record symptoms before and after the procedure in a diary in order to facilitate information transfer. Care logistics responsibilities encapsulate the management aspect of caregiver-companionship.

Care spaces

Caregiver-companions provide informal care to medical tourists in a variety of spaces throughout the medical tourism experience. Our interviews revealed three key spaces of care: travel spaces, the hospital room, and tourist sites in the destination city. Travel spaces are associated with patient transportation to and from the destination country. These include cars, taxis, airports and airplanes. Within such spaces, caregiver-companions provide mobile care, the tasks of which heavily involve hands-on care and logistical care responsibilities. A participant described her caregiver-companion's role while driving home from destination country: 'We needed to stop every hour every half an hour so I could walk . . . to [reduce the] risk [of] blood clots, [and they made] sure I drank water.' Compared to typical patients, medical tourists spend a significant amount of time in travel spaces, and therefore, mobile care is an important contribution offered almost uniquely by caregiver-companions.

Participants commonly recalled the care provided for them by their caregiver-companions within hospital rooms, often discussing it in detail during the interviews. In hospital rooms, caregiver-companions provide patients with care in close proximity because the room is typically a small space where the patient and the caregiver-companion are together almost continuously, day and night, during the hospitalization period. 'My husband stayed in the room with me the whole time,' explained one participant. In hospital rooms, caregiver-companions take on emotional and hands-on care responsibilities. The period of hospitalization was reported to be the most intense period of caregiving because of the near constant proximity between caregiver and patient and continuous care needs. After medical tourists have been released from the hospital, they typically recover in a hotel or resort and have access to travel sites within the destination city. These sites can include markets, recreation venues, public parks and museums. At such travel sites, caregiver-companions provide care and companionship to patients, in part as a way to mitigate the feelings of foreignness and lack of familiarity that are commonly experienced. They provide support by acting as navigators in the city and sometimes by keeping patients from leaving their rooms altogether. A participant recounted how her caregiver-companion took charge at travel sites: 'We wanted to go shopping [and] that's when my husband took over. . . . He knew . . . the exchange [rate], he knew if he was going to be ripped off or not, and he's very good at bartering.' Although the participants we spoke with did not heavily engage in tourist activities, in many cases providing care at those travel sites that were visited enabled the medical tourists to access otherwise inaccessible spaces within the destination city. Caregiver-companions sometimes engaged in tourist activities independently, which some participants thought was a meaningful outlet for stress reduction.

Negotiating relationships

The interviews revealed that throughout the medical tourism experience, relationships among various parties are continuously negotiated. The first relationship to be negotiated is between the medical tourist and a prospective caregiver-companion. Before friends and family become caregiver-companions, medical tourists must first identify their desire to travel abroad with someone rather than alone. The decision to travel accompanied is the first step in transforming the relationship between medical tourists and their friends and family. This decision is sometimes met with ambivalence because they desire support – which they would obtain with a caregiver-companion – but they also often desire independence – which they would maintain if they travelled alone. Many participants reported that caregiver-companions were indispensable, although a good replacement can be found for a higher price in hired private support staff who can provide many of the same caregiving responsibilities while abroad. Generally, 'it's really just good to have someone [you know] there', according to participants who reported feeling less alone in unfamiliar spaces because they were with a familiar person and appreciated the ability to socialize in person and engage in activities with someone they trusted. With this in mind, participants listed a variety of factors that might influence the decision to turn a friend or family member into a caregiver-companion: one's level of personal independence, the extent to which one enjoys the company of others, the invasiveness of the procedure, the level of urgency for the procedure and the types and duration of care likely needed. The decision to travel alone or accompanied was an important one for the medical tourists, and many factors were considered during this relationship negotiation.

Although most participants noted that their caregiver-companions took on a variety of roles and responsibilities, the care offered by each was distinct and determined a great deal by their existing relationship with the medical tourist. The interviews revealed that caregiver-companions tended to mirror the roles and responsibilities they took on in the ongoing relationship they had with the medical tourist under normal circumstances: 'She does that [provide care] with me around [at home] let alone if I was in Chennai, so that's no difference there,' explained one participant. Married and unmarried partners tended to be the most involved caregivers, and among both groups, female partners took on the most hands-on care responsibilities while male partners took leadership over setting agendas, coordinating care and identifying tourism opportunities. Meanwhile, friends tended to offer well-rounded help and cover both emotional and hands-on care, but were less involved overall. Caregiver-companion roles and responsibilities are inconsistent across individuals, but patterns did emerge when the relationship type with the medical tourist was considered.

The typical caregiver-companion experience involved working with formal health care providers at destination facilities towards the provision of patient care. Negotiating relationships with formal providers, as collaborators in caregiving or otherwise, was therefore a necessary role for the companions. Caregiver-companions often did so by taking on some of the patient's communication responsibilities, such as inquiring about the procedure in varying degrees of detail by 'ask[ing] questions while [the patient is] having treatments or while [the formal healthcare providers are] doing tests', providing information the patient forgot and making decisions on the patient's behalf (e.g., meal preferences) when necessary or when asked to do so. They also advocated for the patient to obtain appropriate food, mediated the sometimes overly helpful nature of the nurses and the staff, and defended patients' statements about pain and symptoms. The relationship between health care providers and caregiver-companions tended to be a positive one. Health care providers were reported to be friendly, knowledgeable, and patient, and offered their services to both the patients and the caregiver-companions. Participants noted that health care providers ensured that caregiver-companions were informed and reassured throughout their stay: 'The doctor met with them and told them . . . what to expect. He gave them regular updates while the surgery was going on.' Some caregiver-companions expanded their relationship with health care providers by befriending nurses or by becoming patients themselves after being offered services such as an inexpensive physical exam.

Reflecting on our secondary analysis

Secondary analyses use data from a pre-existing study to answer research questions that are different from those originally posed (Hinds, et al., 1997; Long-Sutehall, et al., 2010; Yardley, et al., 2014). In our case we used data from a study of Canadian medical tourists' decision-making processes to pursue a thematic analysis about caregiver-companions' involvement in this process. Secondary analysis is well-established in quantitative studies, but this is not the case in qualitative research due to several challenges, such as whether data generated by specific qualitative methods is amenable to secondary analysis given how the researcher's understanding may develop through data collection, and whether the extent to which research goals differ between primary and secondary analyses renders them problematic (Hinds, et al., 1997; Thorne, 1998; Heaton, 2008). Heaton (2008) called this the 'problem of data fit'; for example, data collected with one goal in mind may not be appropriate for addressing the goal of an unrelated project. If researchers' original study design and questions reveal new insights, but fail to ask about them specifically, they may impair the data's ability to 'speak' to that aspect of the participant's experience. In addition, Thorne (1998) adds that there may be problems with misrepresentation of participant voices when the analyst is distant from the original study context, as could occur in secondary analyses. Despite these challenges, qualitative researchers such as ourselves are increasingly interested in carrying out secondary analyses as qualitative data sets emerge from labor-intensive processes and often provide highly explanatory data (Long-Sutehall, et al., 2010; Yardley, et al., 2014). To date, because informal caregiving is largely unconsidered in the medical tourism literature, this secondary analysis has proven to be a useful approach to exploring what Long-Sutehall, et al. (2010) term an 'elusive population'. Caregiver-companions are 'elusive' because they may not identify as such, because there is no developed terminology around this emerging phenomenon and because finding medical tourists themselves is already difficult.

In addition to drawing our attention to a thus far unconsidered stakeholder in medical tourism, our secondary analysis has emphasized the importance of relationships between medical tourists and their informal caregivers. Most health geographers' research on informal caregiving has focused on the production of spaces in which care takes place and on care relationships across space (Milligan and Wiles, 2010; Wiles, 2011). Our study findings highlight the roles taken on by caregiver-companions in medical tourism and how their caring relationship with medical tourists is impacted by mobility and the movement between places at an international scale. As noted above, the majority of participants in this analysis were cared for by their spouse, and their relationship at home seemed to predict caregiving behavior while abroad. However, the specific challenges associated with the travel context are somewhat unique to relationships of care in medical tourism. This secondary analysis provides us with a foundation of knowledge about how caring relationships in medical tourism are enacted across space, the mobile nature of this care, and the caring spaces in which it occurs in the medical tourism industry.

'Landscapes of care' is another key concept in qualitative health geography, and this secondary analysis adds to the existing literature by providing new insights into the mobile nature of care provided in medical tourism. Milligan and Wiles (2010) describe how caring relationships engender particular spatialities, noting that we must consider the geographical, social, economic, structural and temporal processes that shape how care is provided and experienced. Qualitative geographical analyses have focused on different scales of care, from the home to the hospital, local to the global, and public to private (Milligan and Wiles, 2010). In the example of caregiver-companions in medical tourism, both care provider and care recipient cross significant physical distances. They are highly mobile, and the landscapes of care in which they operate are foreign, privately-run and likely unfamiliar. The inherent mobility of medical tourism and the different sites of care (i.e., in transit, destination hospitals, recovery spaces and tourist sites) clearly impact the experiences of caregiver-companions, and insights about this practice can contribute to research on other globalized 'landscapes of care' that share elements of travel, such as international retirement migration, physician voluntourism and health worker migration.

This secondary analysis inherits the study limitations in telephone interview-based research generally: lack of visual cues, difficulty establishing rapport, distractions in the interviewees' environment, scheduling across time zones and difficulty communicating due to language barriers or other communication challenges (Hughes, 2008). In addition, our focus on caregiver-companions is limited by the fact that interviews were with medical tourists rather than caregiver-companions themselves. However, this secondary analysis has identified these individuals as pivotal in this health care domain and has promoted their experiences as a future research focus, which is a significant strength.

Acknowledgements

This research was funded by a Catalyst Grant awarded by the Canadian Institutes of Health Research. Valorie Crooks is funded by a Scholar Award from the Michael Smith Foundation for Health Research and holds the Canada Research Chair in Health Service Geographies.

References

Armstrong, P., Armstrong, H. and Coburn, D. (eds.), 2001. *Unhealthy times: political economy perspectives on health and care*. London, UK: Oxford University Press.

Bertinato, L., Busse, R., Fahy, N., Legido-Quigley, H., McKee, M., Palm, W., Passarani, I., and Ronfini, F., 2005. *Policy brief: cross-border health care in Europe*. European Observatory on Health Systems and Policies: World Health Organization.

Botten, G., Grepperud, S., and Nerland, S., 2004. Trading patients: lessons from Scandinavia. *Health Policy*, 69, pp. 317–327.

Casey, V., Crooks, V.A., Snyder, J., and Turner, L., 2013a. 'You're dealing with an emotionally charged individual . . . ': an industry perspective on the challenges posed by medical tourists' informal caregiver-companions. *Globalization and Health*, (9)31.

Casey, V., Crooks, V.A., Snyder, J., and Turner, L., 2013b. Knowledge brokers, companions, and navigators: qualitatively examining informal caregivers' roles in medical tourism. *International Journal for Equity in Health*, (12)94.

CBC, 2010. MS patients rally to get treatment covered. *CBC News*, Available at: http://www.cbc.ca/news/canada/calgary/ms-patients-rally-to-get-treatment-covered-1.892571. Accessed August 15, 2014.

Connell, J., 2008. Tummy Tucks and the Taj Mahal? Medical tourism and the globalization of health care. In Woodside, A. G., and Martin, D. (eds.) *Tourism management: analysis, behaviour and strategy*. Wallingford: CAB International. pp. 232–244.
Cortez, N., 2008. Patients without borders: the emerging global market for patients and the evolution of modern health care. *Indiana Law Journal*, 83(1), pp. 71–132.
Crooks, V. A., Cameron, K., Chouinard, V., Johnston, R., Snyder, J., and Casey, V., 2012. Use of medical tourism for hip and knee surgery in osteoarthritis: a qualitative examination of distinctive attitudinal characteristics among Canadian patients. *BMC Health Services Research*, 12(417).
Crooks, V. A., Kingsbury, P., Snyder, J., and Johnston, R., 2010. What is known about the patient's experience of medical tourism? A scoping review. *BMC Health Services Research*, 10(266).
Ehrbeck, T., Guevara, C., Mango, P.D., Cordina, R., and Singhal, S., 2008. Health care and the consumer. *McKinsey Quarterly*, 4, pp. 80–91.
Fowlie, J., 2010. B.C.: come for the great outdoors, and surgery; medical tourism. *The National Post*, March 10. p. A2.
Garcia-Altes, A., 2005. The development of health tourism services. *Annals of Tourism Research*, 32(1), pp. 262–266.
Garud, A. D., 2005. Medical tourism and its impact on our healthcare. *National Medical Journal of India*, 18, pp. 318–319.
Heaton, J., 2008. Secondary analysis of qualitative data: an overview. *Historical Social Research*, 33(3), pp. 33–45.
Hinds, P. S., Vogel, R. J., and Clarke-Steffen, L., 1997. The possibilities and pitfalls of doing a secondary analysis of a qualitative data set. *Qualitative Health Research*, 7(3), pp. 408–424.
Hopkins, L., Labonte, R., Runnels, V., and Packer, C., 2010. Medical tourism today: what is the state of existing knowledge? *Journal of Public Health Policy*, 31(2), pp. 185–198.
Hughes, R., 2008. Telephone interview. In Given, L. (ed.), 2008. *The SAGE encyclopedia of qualitative research methods*. Thousand Oaks: SAGE. pp. 863–864.
Irwin, S., 2013. Qualitative secondary data analysis: ethics, epistemology and context. *Progress in Development Studies*, 13(4), pp. 295–306.
Johnston, R., Crooks, V. A. and Snyder, J., 2012. 'I didn't even know what I was looking for': A qualitative study of the decision-making processes of Canadian medical tourists. *Globalization and Health*, 8(23).
Laidlaw, S., 2008. Global health care on its way. *Toronto Star*, May 31. p. L1.
Lautier, M., 2008. Export of health services from developing countries: the case of Tunisia. *Social Science and Medicine*, 67(1), pp. 101–110.
Leahy, A., 2008. Medical tourism: the impact of travel to foreign countries for healthcare. *The Surgeon*, 6(5), pp. 260–261.
Long-Sutehall, T., Sque, M., and Addington-Hall, J., 2010. Secondary analysis of qualitative data: a valuable method for exploring sensitive issues with an elusive population? *Journal of Research in Nursing*, 16(4), pp. 335–344.
Milligan, C. and Wiles, J., 2010. Landscapes of care. *Progress in Human Geography*, 34(6), pp. 736–754.
Mudur, G., 2004. Hospitals in India woo foreign patients. *British Medical Journal*, 328(7452), pp. 1338.
Pachanee, C., and Wibulpolprasert, S., 2006. Incoherent policies on universal coverage of health insurance and promotion of international trade in health services in Thailand. *Health Policy Plan*, 21(4), pp. 310–318.

Sen Gupta, A., 2008. Medical tourism in India: winners and losers. *Indian Journal of Medical Ethics*, 5(1), pp. 4–5.

Snyder, J., Crooks, V. A., and Johnston, R., 2012. Perceptions of the ethics of medical tourism by Canadian patients: comparing patient and academic perspectives. *Public Health Ethics*, 5(1), pp. 38–46.

Thorne, S., 1998. Ethical and representational issues in qualitative secondary analysis. *Qualitative Health Research*, 8(4), pp. 547–55.

Weber, R., 2010. Health Tourism on BC Government's Agenda. *The Globe and Mail*, March 15. p. D3.

Westhead, R., 2010. 'Superbug' threat doesn't scare Canadian medical tourists in India. *The Toronto Star*, August 16. p. H02.

White, N., 2010. Superbug detected in GTA. *The Toronto Star*, August 21. p. H03.

Whittaker, A., 2008. Pleasure and pain: medical travel in Asia. *Global Public Health*, 3(3), pp. 271–290.

Wibulpolprasert, S., Pachanee, C., Pityarangsarit, S., and Hempisut, P., 2004. International service trade and its implications for human resources for health: a case of Thailand. *Human Resources for Health*, 2, p.10.

Wiles, J., 2011. Reflections on being a recipient of care: vexing the concept of vulnerability. *Social and Cultural Geography*, 12(6), pp. 573–588.

Yardley, S. J., Watts, K. M., Pearson, J., and Richardson, J. C., 2014. Ethical issues in the reuse of qualitative data: Perspectives from literature, practice and participants. *Qualitative Health Research*, 24(1), pp. 102–113.

Yelaja, P., 2006. India offers surgery in a hurry. *Toronto Star*, June 17. p. A1.

York, D., 2008. Medical tourism: the trend toward outsourcing medical procedures to foreign countries. *Journal of Continuing Education in the Health Professions*, 28(2), pp. 99–102.

15 Conclusion

Robin Kearns

Reflections on praxis

I am writing from hospital, where I have accompanied a family member onto an assessment ward. We are on the eighth floor, waiting. Being patient in the course of one of us becoming a patient. Mercifully, there is the view from the expansive windows out over the city and to the islands in the gulf. The green of the grass and blue of the sea and sky speak in colors of a world beyond, the world of another day, a tomorrow when all will be well. The small pieces of artwork are reminders of other worlds too, miniature suggestions of imagined spaces outside these grey-green walls and floors, stark white sheets and a clock running forty minutes late. Drawers by the bedside table have faded labels that leave one guessing. One says 'vomit cartons'; another 'blue sheets'. As my loved one sleeps, my eyes glance up to the door in startled anticipation each time someone walks past. Invariably they walk on, and we wait on for a doctor to visit. Rendered still by the unexpected circumstances of the day, I cannot help but observe the details of this place. As time hangs heavy and the patient sleeps, my thoughts turn to writing this chapter, and I recall two other memorable times of waiting in that formative first year as a researcher after completing my PhD.

I was interviewing community-based patients who had formerly been residents of Auckland's large and recently closed psychiatric hospital. In accordance with the hospital board's ethical guidelines, nurses established the willingness of potential participants and passed their details onto me. When it came time for me to contact Beth, she asked to meet me at work. She was employed at 'Val's Place' on K Road. From the street name, I immediately surmised it was part of the sex industry for which that street was notorious. I felt uneasy. To interview or not to interview, that was the question. In fidelity to the research design and Beth's willingness to talk, I gathered the courage to go to her workplace by asking my partner to come along. Neither of us has stepped into such a place before. We were met with brusque curiosity by the respectable-looking woman at reception and told we would find Beth in the parlor. In the low light she came towards us hesitatingly and led us up into an equally low-lit room furnished with bed, bath, whips and skulls with protruding incense sticks. I proceeded to ask her to sign a consent form, and she reluctantly did so but asked why I had brought someone else.

That, she said, would make it seem like we were official inspectors of some sort. I sat, gingerly perched on the edge of the bed, asking what suddenly seemed to be very banal questions while large beads of sweat rolled down my back. Most of the girls here are from the hospital, she said. As I descended the stairs having omitted some of the questions on my list, I wondered to what extent I had made things more difficult for her by bringing someone along whose company was to make me a degree more comfortable.

Later that year I was seeking to understand the social significance of rural health clinics in the far north of New Zealand. There, my embodiment was central to the construction of knowledge. Within the waiting area of a particularly memorable clinic, I positioned myself in an attempt to seem neither out-of-place (and thus inhibit conversation by gazing at others) nor be overly in-place (and hence welcoming of engagement with others). My compromise involved the daily newspaper and pretending to do the crossword puzzle. I was naïve to think this could be a disguise for my attempt at partial participant observation. Soon two 'locals' offered to help with the puzzle I was half-heartedly completing. Later, an elder entered and proceeded to kiss and welcome (in *te reo* – Maori language) all present, myself included. I was clearly in 'her' clinic, not just any public space. There was nowhere to hide. Our bodies connected in an embrace, rendering foolish my prior thoughts of watching community life from a corner. The binary construct of researcher and researched was rendered permeable and re-embodied within a web of sociocultural relations.

Health geography: 'more-than-'

In a sense, writing a conclusion chapter to this book implies more completeness and closure than is warranted for a set of essays on innovative qualitative methods in health geography. The foregoing chapters have shown that health geography is increasingly animated with the art of the possible, infused with the spirit of 'more-than-' (human, material, medical?) that has pervaded recent human geography. This book has illustrated how much more health geography is and can be.

The editors emphasized in the introduction that this is less a 'how-to' book, and more a 'how-to-learn-from' collection. Indeed, the book's opening sentence stresses the *doing* of research. Each chapter's inclusion of a section on 'praxis' has reinforced this emphasis on the practical with candid reflections on the process and progress of each author's (or set of authors') expression of health geography in action. The editors see praxis as focusing on the spaces in between the 'doing of research'. Contributors' chapters have offered illuminating windows into the worlds of geographers actively listening to their participants, their worlds and each other as they take stock during or after immersion in the field experience.

The other word stressed at the outset of this collection is 'novelty'. The shift from predominantly thinking medically to thinking about health by geographers has involved, among other things, something of a glasnost in terms of research design. A quarter of a century ago the subdiscipline existed under a largely medicalized guise and largely operated in grip of an enumerative orthodoxy.

Today, data no longer need to be an entity 'out there' that one obtains and analyses. Rather, through a focus on the links between health and place, geographers have increasingly placed themselves more self-consciously in the field of inquiry. First, through considering the implications of their place within research relations, there has been an elevated recognition of the ways that research engagements invariably involve the co-production of relationships and meanings that literally take place. Second, in the spirit of Nast's (1994) proclamation that 'we are always and everywhere in the field', health geographers have acknowledged that the field is not necessarily a site to be journeyed to, but rather can be the sonic, tactile and thoroughly peopled landscape of everyday life. This book is ample evidence that Thrift's (2000) urging for qualitative researchers to be more methodologically imaginative has found definite traction within health geography. As its contributors attest, new approaches have been emphasized and discussed with gusto by practitioners inspired by a potent mix of theory and their own creativity.

The foregoing chapters tacitly embrace earlier calls for more sensuous, embodied and emotional geographies (Longhurst, et al., 2008; Davidson and Milligan, 2004; Rodaway, 2002). Chapters have literally walked the talk, exploring mobile methods and, for instance, the old adage that a picture can paint a thousand words (in chapters dealing with drawing and photo elicitation). While such adventurous approaches risk taking us out of our comfort zones in terms of fieldwork and analysis, the opportunities that lie within them involve deeper encounters with the humanity and vulnerability of those we engage. Many of the present contributions suggest something of the 'witnessing' that Dewsbury (2003) identifies as a disposition involving critical awareness of our complicity in the co-production of our world. This term seems useful for its sense of engagement in the time and place of our co-dwelling and encounter, for to be a witness is to be involved by virtue of one's presence and not just observing or counting remotely.

Becker (1965, p. 602) writes that, 'no matter how carefully one plans in advance, research is designed in the course of its execution'. That being the case, this collection serves a critically important role in challenging health geographers to offer 'witness' to their research involvements, to remain acutely aware of unexpected developments in the course of their research through maintaining a disposition of attentiveness, keeping field notes, and being flexible in order to accommodate improvised adaptations in practice throughout the research process. This collection is novel for its level of frankness among health geographers: a collective admission of uncertainties and self-doubt yet unexpected insights. Through this admission of the 'warts and all' of qualitative research, we have seen in its chapters the emergence of more deeply *human* health geographies.

If the semi-structured interview held sway as the method of choice in the first decades of the 'qualitative turn', there has been an increasing dissatisfaction with its capacity to yield information that satisfactorily 'speaks to' what is *really* going on in the world. To complement the interview, as we have seen in this collection, there have been attempts to achieve a greater transparency of self on the part of participants, researchers or both. Yet strangely, perhaps, given the passage of time since the 'qualitative turn' became apparent, there remains a long shadow of suspicion of qualitative research in human geography. Recently, by way of example, I assigned to a postgraduate class an article published in *Health and Place* that drew on in-depth interviews with six participants. In their response papers many in the class criticized the authors for recruiting an unrepresentative sample that could not be statistically significant.

A key challenge, if not constraint, to responsive and evolutionary field method is the increasingly rule-bound systems which govern how university researchers conduct themselves in the field. A hypervigilance to scrutinizing research plans, but an under-attention to subsequent field relations can leave an asymmetry. One can be tied, through consent agreements, to a prescriptive course of action, yet at liberty to engage or not with participants or a community-at-large on the far side of completion. What Haggarty (2004) has called 'ethics creep' has resulted in students frequently talking in terms of 'getting ethics'. Thus the process of credentialing has often become more a matter of sanctioning plans than an awareness of conduct throughout. Looking back on my own process of noting conversations in rural clinic waiting rooms (see the praxis section above), I was told years later in no uncertain terms that such practice would not have gained approval from the university ethics committee had one existed at the time. This is ironic, perhaps, as the resulting paper was used by the local community to save its demise under health care restructuring (Kearns 1991).

Qualitative methods have become the new orthodoxy in human geography at large, witnessed for instance by a new fourth and expanded edition of *Qualitative Research Methods in Human Geography* (2010) going to press. What are our responsibilities in this time when what was marginal is now the norm? First, even if we are practitioners who prefer to grapple with social situations more than statistical significance, we must surely encourage the continued numeracy of students and remain open to conversations – if not collaboration – with more quantitatively inclined colleagues. Second, there is a need to ensure that the quest for creativity in health geography methodologies is not pursued at the cost of losing rigor and a connectedness between theory and practice. As Baxter and Eyles (1997) pointed out, there can be tension between being creative in the qualitative research process and the quest to assess its rigor, a process that implies the use of standardized procedures and styles of reporting.

One of the key challenges for qualitative researchers is succeeding in being funded by agencies that increasingly seek clearly articulated and predicted outcomes (Crang, 2002). A commitment to evolving one's research program in response to the relational dynamics encountered and the situations witnessed remains an obstacle for a bureaucratized world of outputs, objectives and milestones. As some of the chapters in this volume suggest, perhaps there is a responsibility of witnessing in writing to the nuanced insights gained in praxis in addition to and after the business of reporting and meeting of objectives. This could be a corrective to what Crang (2002) observes: much qualitative research in human geography having an 'evidential realist flavour'. By this he means that notwithstanding considerable discussion about and reflection on positionality, the analysis is conservative and pragmatic for its selected use of evidence in an attempt to verify the researchers' initial contentions. There is clearly a balance needed here between creativity and conformity. As Parr (2004) argues, for instance, there is a danger of research becoming overly subjective, thus rendering one's writing unhelpful for promoting changes in policy and contributing to emancipatory politics.

A good point on which to close this volume aligns with one made in Chapter 1: that it is an exciting time to be practicing qualitative research in health geography. Two decades ago, a call to respond to diversity and see a closer engagement with place through health (Kearns, 1993) was rejected as 'misinformed' and 'quite wrong' (Meade and Mayer, 1994). Today there are fewer such disciplining voices calling youthful enthusiasm into line. Rather, in these more liberal (as well as neoliberal) times, there is an acknowledged diversity of ways to address the subtle dimensions of understanding health issues in the context of fieldwork. We have become, in the words of Kearns and Moon (2002), a magpie discipline (at times playfully) accumulating materials from various sources that can be reassembled, within a backdrop of theory, in the cause of interpretation and explanation.

The diversity of the research included in this book is testament to the courage of authors and the encouragement of supervisors as a new generation of postgraduate students have pioneered new ways of seeing and doing. As a collective journey, it has included candid autobiography, elicited drawing and narrative, the use of video and walking-with-talking encounters through to engagement with music. In sum, these chapters offer to move us away from presuming to understand the world simply through what is said in a standard interview.

Health geographers have increasingly made space for difference and acknowledged interconnecting webs of relationships. Within these webs, no aspect can be fully understood without considering the whole. The socioecological model of health that inspired much early work on health and place symbolically expressed this connectedness through being, symbolically, an 'echo-system' in which disruptions to everyday experience reverberated around domains of experience. This way of seeing health as disaggregated into aspects of the environment helped geographers move away from under the shadow of medicine. However, it is less well equipped to assist us to expand our horizons into the embrace of well-being, an even more nuanced concept than health (Fleuret and Atkinson, 2007). Rather it is more likely that a version of Deleuzian thinking will assist geographical research, given the indeterminacy of feelings and experiences associated with health and well-being. For an assemblage perspective offers a way of talking about the social world that removes us from the presuppositions we often hold. Instead of discrete social objects or things (clinics, doctors, pills) we are drawn to think in terms of a patchwork and a transitory world of fluidity and variable configurations.

As we look beyond this book, health geography is set to benefit from the methodological enrichment of these chapters as it becomes less a literal (sub) discipline and more a domain of disciplinary endeavor. The challenge on the journey will be to balance creative engagement with an ongoing commitment to advancing human health and well-being.

References

Baxter, J., and Eyles, J., 1997. Evaluating qualitative research in social geography: establishing 'rigour' in interview analysis. *Transactions of the Institute of British Geographers*, 22 (4), pp. 505–525.

Becker, Howard S., 1965. Review of *Sociologists at work: essays on the craft of social research* by Philip E. Hammond. *American Sociological* Review 30(4), pp. 602–603.

Crang, M., 2002. Qualitative methods; the new orthodoxy? *Progress in Human Geography*, 26(5), pp. 647–655.

Davidson, J., and Milligan, C. 2004. Embodying emotion sensing space: introducing emotional geographies. *Social & Cultural Geography*, 5(4), pp. 523–532.

Dewsbury, J. D., 2003: Witnessing space: knowledge without contemplation'. *Environment and Planning A*, 35, pp. 1907–32.

Fleuret, S., Atkinson, S., 2007. Wellbeing, health and geography: a critical review and research agenda. *New Zealand Geographer*, 63(2), pp. 106–118.

Haggarty, K., 2004. Ethics creep: governing social science research in the name of ethics. *Qualitative Sociology* 27(4), pp. 391–414.

Hay, I. (ed.), 2010. *Qualitative Research Methods in Human Geography*. 3rd ed. Melbourne: Oxford University Press.

Kearns, R. A., 1991. The place of health in the health of place: the case of the Hokianga special medical area. *Social Science and Medicine*, 33, pp. 519–530.

Kearns, R. A., 1993. Place and health: towards a reformed medical geography. *The Professional Geographer*, 45, pp. 139–147.

Longhurst, R., Ho, E., and Johnston, L., 2008. Using 'the body' as an 'instrument of research': kimch'i and pavlova. *Area*, 40(2), pp. 208–217.

Mayer, J. D., and Meade, M. S., 1994. A Reformed medical geography reconsidered, *The Professional Geographer*, 46(1), pp. 103–106.

Parr, H., 2004. Medical geography: critical medical and health geography? *Progress in Human Geograohy*, 28(2), pp. 246–257.

Rodaway, P., 2002. *Sensuous geographies: body, sense and place*. London: Routledge.

Thrift, N., 2000. Dead or alive? In Cook, I., Crouch, D., Naylor, S., Ryan, J. (eds.) *Cultural turns /geographical turns: perspectives on cultural geography*. Harlow, Essex: Prentice Hall.

Index